Compiled by Chinese Academy of Tropical Agricultural Sciences (CATAS) and
Chinese Society for Tropical Crops (CSTC)
A Series of Books for Field Guide to Common Plants in FSM

General Editor: Liu Guodao

Coconut Germplasm Resources in the Federated States of Micronesia

Editors in Chief: Fan Haikuo Gong Shufang Wang Yuanyuan

China Agricultural Science and Technology Press

图书在版编目（CIP）数据

密克罗尼西亚联邦椰子种质资源图鉴 = Coconut Germplasm Resources in the Federated States of Micronesia / 范海阔，弓淑芳，王媛媛主编 . —北京：中国农业科学技术出版社，2021.5

（密克罗尼西亚常见植物图鉴系列丛书 / 刘国道主编）

ISBN 978-7-5116-5290-4

Ⅰ . ①密… Ⅱ . ①范… ②弓… ③王… Ⅲ . ①椰子—种质资源—密克罗尼西亚联邦—图集 Ⅳ . ① S667.402.4-64

中国版本图书馆 CIP 数据核字（2021）第 068611 号

责任编辑　徐定娜
责任校对　贾海霞
责任印制　姜义伟　王思文

出 版 者	中国农业科学技术出版社
	北京市中关村南大街 12 号　邮编：100081
电　　话	（010）82109707（编辑室）　（010）82109702（发行部）
	（010）82109709（读者服务部）
传　　真	（010）82109707
网　　址	http://www.castp.cn
发　　行	各地新华书店
印 刷 者	北京科信印刷有限公司
开　　本	787 mm×1 092 mm　1/16
印　　张	4
字　　数	162 千字
版　　次	2021 年 5 月第 1 版　2021 年 5 月第 1 次印刷
定　　价	108.00 元

◆版权所有·侵权必究◆

About the Author

Dr. Liu Guodao, born in June 1963 in Tengchong City, Yunnan province, is the incumbent Vice President of Chinese Academy of Tropical Agricultural Sciences (CATAS). Being a professor and PhD tutor, he also serves as the Director-General of the China-Republic of the Congo Agricultural Technology Demonstration Center.

In 2007, he was granted with his PhD degree from the South China University of Tropical Agriculture, majoring in Crop Cultivation and Farming.

Apart from focusing on the work of CATAS, he also acts as a tutor of PhD candidates at Hainan University, Member of the Steering Committee of the FAO Tropical Agriculture Platform (TAP), Council Member of the International Rubber Research and Development Board (IRRDB), Chairman of the Chinese Society for Tropical Crops, Chairman of the Botanical Society of Hainan, Executive Director of the Chinese Grassland Society and Deputy Director of the National Committee for the Examination and Approval of Forage Varieties and the National Committee for the Examination and Approval of Tropical Crop Varieties.

He has long been engaged in the research of tropical forage. He has presided over 30 national, provincial and ministerial-level projects: namely the "National Project on Key Basic Research (973 Program)" and international cooperation projects of the Ministry of Science and Technology, projects of the National Natural Science Foundation of China, projects of the International Center for Tropical Agriculture in Colombia and a bunch of projects sponsored by the Ministry of Agriculture and Rural Affairs (MARA) including the Talent Support Project, the "948" Program and the Infrastructure Project and Special

Scientific Research Projects of Public Welfare Industry.

He has published more than 300 monographs in domestic and international journals such as "New Phytologist" "Journal of Experimental Botany" "The Rangeland Journal" "Acta Prataculturae Sinica" "Acta Agrestia Sinica" "Chinese Journal of Tropical Crops", among which there nearly 20 were being included in the SCI database. Besides, he has compiled over 10 monographs, encompassing "Poaceae Plants in Hainan" "Cyperaceae Plants in Hainan" "Forage Plants in Hainan" "Germplasm Resources of Tropical Crops" "Germplasm Resources of Tropical Forage Plants" "Seeds of Tropical Forage Plants" "Chinese Tropical Forage Plant Resources". As the chief editor, he came out a textbook-*Tropical Forage Cultivation*, and two series of books-*Practical Techniques for Animal Husbandry in South China Agricultural Regions* (19 volumes) and *Practical Techniques for Chinese Tropical Agriculture "Going Global"* (16 volumes).

He has won more than 20 provincial-level and ministerial-level science and technology awards. They are the Team Award, the Popular Science Award and the First Prize of the MARA China Agricultural Science and Technology Award, the Special Prize of Hainan Natural Science Award, the First Prize of the Hainan Science and Technology Progress Award and the First Prize of Hainan Science and Technology Achievement Transformation Award.

He developed 23 new forage varieties including Reyan No. 4 King grass. He was granted with 6 patents of invention and 10 utility models by national patent authorities. He is an Outstanding Contributor in Hainan province and a Special Government Allowance Expert of the State Council.

Below are the awards he has won over the years: in 2020, "the Ho Leung Ho Lee Foundation Award for Science and Technology Innovation"; in 2018, "the High-Level Talent of Hainan province" "the Outstanding Talent of Hainan province" "the Hainan Science and Technology Figure"; in 2015, Team Award of "the China Agricultural Science and Technology Award" by the Ministry of Agriculture; in 2012, "the National Outstanding Agricultural Talents Prize" awarded by the Ministry of Agriculture and as team leader of the team award: "the Ministry of Agriculture Innovation Team" (focusing on the research of Tropical forage germplasm innovation and utilization); in 2010, the first-level candidate of the "515 Talent Project" in Hainan province; in 2005, "the Outstanding Talent of Hainan province"; in 2004, the first group of national-level candidates for the "New Century Talents Project" "the 4th Hainan Youth Science and Technology Award" "the 4th Hainan Youth May 4th Medal" "the 8th China Youth Science and Technology Award" "the Hainan Provincial International Science and Technology Cooperation Contribution Award"; in 2003, "a Cross-Century Outstanding Talent" awarded by the Ministry of Education; In 2001, "the 7th China Youth Science and Technology Award" of Chinese Association of Agricultural Science Societies, "the National Advanced Worker of Agricultural Science and Technology"; in 1993, "the Award for Talents with Outstanding Contributions after Returning to China" by the State Administration of Foreign Experts Affairs.

 Dr. Fan Haikuo, born in 1976, graduated from Sichuan Agricultural University, majoring in pomology. He has been rated as the "Nanhai Expert" and "Leading Talent" of Hainan, mainly engaged in coconut germplasm resources and breeding research. He serves as the Deputy Director of the Coconut Research Institute of Chinese Academy of Tropical Agricultural Sciences, Expert Member of the National Variety Approval Committee of the Ministry of Agriculture and Rural Affairs, the Secretary-General of the Hainan Committee on Standard Systems for Coconut and Betel Nut Industries, and the Hainan Coconut Industry Association. He has undertaken 44 provincial and ministerial-level scientific research projects, obtained 12 patents, enacted 5 industrial standards, cultivated 5 new varieties, published 12 monographs, and 76 papers, 12 of which have been included in SCI database.

A Series of Books for Field Guide to Common Plants in FSM

General Editor: Liu Guodao

Coconut Germplasm Resources in the Federated States of Micronesia Editorial Board

Editors in Chief:

Fan Haikuo   Gong Shufang   Wang Yuanyuan

Associate editors in chief:

Dong Dingchao   Yang Hubiao   Zheng Xiaowei   Tang Qinghua

Zhang Zhaohua

Members (in alphabet order of surname):

Chen Gang   Dong Dingchao   Fan Haikuo   Gong Shufang

Hao Chaoyun   Huang Guixiu   Li Weiming   Liu Guodao

Tang Qinghua   Wang Qinglong   Wang Yuanyuan   Yang Guangsui

Yang Hubiao   Zheng Xiaowei   Zhang Zhaohua

Photographers:

Fan Haikuo   Yang Hubiao

Translator:

Cui Shuang

The President
Palikir, Pohnpei
Federated States of Micronesia

FOREWORD

It is with great pleasure that I present this publication, "Agriculture Guideline Booklet" to the people of the Federated States of Micronesia (FSM).

The Agriculture Guideline Booklet is intended to strengthen the FSM Agriculture Sector by providing farmers and families the latest information that can be used by all in our communities to practice sound agricultural practices and to support and strengthen our local, state and national policies in food security. I am confident that the comprehensive notes, tools and data provided in the guideline booklets will be of great value to our economic development sector.

Special Appreciation is extended to the Government of the People's Republic of China, mostly the Chinese Academy of Tropical Agricultural Sciences (CATAS) for assisting the Government of the FSM especially our sisters' island states in publishing books for agricultural production. Your generous assistance in providing the practical farming techniques in agriculture will make the people of the FSM more agriculturally productive.

I would also like to thank our key staff of the National Government, Department of Resources and Development, the states' agriculture and forestry divisions and all relevant partners and stakeholders for their kind assistance and support extended to the team of Scientists and experts from CATAS during their extensive visit and work done in the FSM in 2018.

We look forward to a mutually beneficial partnership.

Sincerely,

David W. Panuelo
President

Preface

Claiming waters of over 3,000 square kilometers, the vast area where Pacific island countries nestle is home to more than 10,000 islands. Its location at the intersection of the east-west and north-south main traffic artery of Pacific wins itself geo-strategic significance. There are rich natural resources such as agricultural and mineral resources, oil and gas here. The relationship between the Federated States of Micronesia (hereinafter referred to as FSM) and China ushered in a new era in 2014 when Xi Jinping, President of China, and the leader of FSM decided to establish a strategic partnership on the basis of mutual respect and common development. Mr Christian, President of FSM, took a successful visit to China in March 2017 during which a consensus had been reached between the leaders that the traditional relationship should be deepened and pragmatic cooperation (especially in agriculture) be strengthened. This visit pointed out the direction for the development of relationship between the two countries. In November 2018, President Xi visited Papua New Guinea and in a collective meeting met 8 leaders of Pacific Island countries (with whom China has established diplomatic relation). China elevated the relationship between the countries into a comprehensive and strategic one on the basis of mutual respect and common development, a sign foreseeing a brand new prospect of cooperation.

The government of China launched a project aimed at assisting FSM in setting up demonstration farms in 1998. Until now, China has completed 10 agricultural technology cooperation projects. To answer the request of the government of FSM, Chinese Academy of Tropical Agricultural Sciences (hereinafter referred to as CATAS), directly affiliated with the

Ministry of Agriculture and Rural Affairs of China, was elected by the government of China to carry out training courses on agricultural technology in FSM during 2017—2018. The fruitful outcome is an output of training 125 agricultural backbone technicians and a series of popular science books which are entitled "Field Guide to Forages in the Federated States of Micronesia" "Field Guide to Flowers and Ornamental Plants in the Federated States of Micronesia" "Field Guide to Medicinal Plants in the Federated States of Micronesia" "Field Guide to Fruits and Vegetables in the Federated States of Micronesia" "Coconut Germplasm Resources in the Federated States of Micronesia" and "Field Guide to Plant Diseases, Insect Pests and Weeds in the Federated States of Micronesia".

In these books, 492 accessions of germplasm resources such as coconut, fruits, vegetables, flowers, forages, medical plants, and pests and weeds are systematically elaborated with profuse inclusion of pictures. They are rare and precious references to the agricultural resources in FSM, as well as a heart-winning project among China's aids to FSM.

Upon the notable moment of China-Pacific Island Countries Agriculture Ministers Meeting, I would like to send my sincere respect and congratulation to the experts of CATAS and friends from FSM who have contributed remarkably to the production of these books. I am firmly convinced that the exchange between the two countries on agriculture, culture and education will be much closer under the background of the publication of these books and Nadi Declaration of China-Pacific Island Countries Agriculture Ministers Meeting, and that more fruitful results will come about. I also believe that the team of experts in tropical agriculture mainly from the CATAS will make a greater contribution to closer connection in agricultural development strategies and plans between China and FSM, and closer cooperation in exchanges and capacity-building of agriculture staffs, in agricultural science and technology for the development of agriculture of both countries, in agricultural investment and trade, in facilitating FSM to expand industry chain and value chain of agriculture, etc.

Qu Dongyu

Director General

Food and Agriculture Organization of the United Nations

July 23, 2019

Located in the northern and central Pacific region, the Federated States of Micronesia (FSM) is an important hub connecting Asia and America. Micronesia has a large sea area, rich marine resources, good ecological environment, and unique traditional culture.

In the past 30 years since the establishment of diplomatic relations between China and FSM, cooperation in diverse fields at various levels has been further developed. Since the 18th National Congress of the Communist Party of China, under the guidance of Xi Jinping's thoughts on diplomacy, China has adhered to the fine diplomatic tradition of treating all countries as equals, adhered to the principle of upholding justice while pursuing shared interests and the principle of sincerity, real results, affinity, and good faith, and made historic achievements in the development of P.R. China-FSM relations.

The Chinese government attaches great importance to P.R. China-FSM relations and always sees FSM as a good friend and a good partner in the Pacific island region. In 2014, President Xi Jinping and the leader of the FSM made the decision to build a strategic partnership featuring mutual respect and common development, opening a new chapter of P.R. China-FSM relations. In 2017, FSM President Peter Christian made a successful visit to China. President Xi Jinping and President Christian reached broad consensuses on deepening the traditional friendship between the two countries and expanding practical cooperation between the two sides, and thus further promoted P.R. China-FSM relations. In 2018, Chinese President Xi Jinping and Micronesian President Peter Christian had a successful meeting again in PNG and made significant achievements, deciding to upgrade P.R. China-FSM relations to a new stage of Comprehensive Strategic Partnership, thus charting the course for future long-term development of P.R. China-FSM relations.

In 1998, the Chinese government implemented the P.R. China-FSM demonstration farm project in FSM. Ten agricultural technology cooperation projects have been completed, which has become the "golden signboard" for China's aid to FSM. From 2017 to 2018, the Chinese Academy of Tropical Agricultural Sciences (CATAS), directly affiliated with the Ministry of Agriculture and Rural Affairs, conducted a month-long technical training on pest control of coconut trees in FSM at the request of the Government of FSM. 125 agricultural managers, technical personnel and growers were trained in Yap, Chuuk, Kosrae and Pohnpei, and the biological control technology demonstration of the major dangerous pest, Coconut Leaf Beetle, was carried out. At the same time, the experts took advantage of the spare time of the training course and spared no effort to carry out the preliminary evaluation of the investigation and utilization of agricultural resources, such as coconut, areca nut, fruit tree, flower, forage, medicinal plant, melon and vegetable, crop disease, insect pest and weed diseases, in the field in conjunction with Department of Resources and Development of FSM and the vast number of trainees, organized and compiled a series of popular science books, such as "Field Guide to Forages in the Federated States of Micronesia" "Field Guide to Flowers and Ornamental Plants in the Federated States of Micronesia" "Field Guide to Medicinal Plants in the Federated States of Micronesia" "Field Guide to Fruits and Vegetables in the Federated States of Micronesia" "Coconut Germplasm Resources in the Federated States of Micronesia" and "Field Guide to Plant Diseases, Insect Pests and Weeds in the Federated States of Micronesia".

The book introduces 37 kinds of coconut germplasm resources, 60 kinds of fruits and vegetables, 91 kinds of angiosperm flowers as well as 13 kinds of ornamental pteridophytes, 100 kinds of forage plants, 117 kinds of medicinal plants, 74 kinds of crop diseases, pests and weed diseases, in an easy-to-understand manner. It is a rare agricultural resource illustration in FSM. This series of books is not only suitable for the scientific and educational workers of FSM, but also it is a valuable reference book for industry managers, students, growers and all other people who are interested in the agricultural resources of FSM.

This series is of great significance for it is published on the occasion of the 30th anniversary of the establishment of diplomatic relations between the People's Republic of

China and FSM. Here, I would like to pay tribute to the experts from CATAS and the friends in FSM who have made outstanding contributions to this series of books. I congratulate and thank all the participants in this series for their hard and excellent work. I firmly believe that based on this series of books, the agricultural and cultural exchanges between China and FSM will get closer with each passing day, and better results will be achieved more quickly. At the same time, I firmly believe that the Chinese Tropical Agricultural Research Team, with CATAS as its main force, will bring new vigour and make new contributions to promoting the in-depth development of the strategic partnership between the People's Republic of China and the Federated States of Micronesia, strengthening solidarity and cooperation between P.R. China and the developing countries, and the P.R. China-FSM joint pursuit of the Belt and Road initiative and building a community with a shared future for the humanity.

Ambassador Extraordinary & Plenipotentiary of
the People's Republic of China to
the Federated States of Micronesia
May 23, 2019

Foreword

Coconut (*Cocos nucifera* L.) is the main woody oil crop and an important food energy crop in the tropics. Coconut is also an important economic crop in the tropics due to its low cultivation cost and high comprehensive utilization value. The Food and Agriculture Organization of the United Nations (FAO) attaches great importance to the coconut industry and believes that coconut cultivation is an important way to solve the needs of people in the tropics for protein, fat and energy, increase employment opportunities for farmers, and help farmers to get rid of poverty.

According to the statistics of the Asia Pacific Coconut Community (APCC), coconut is cultivated currently by 93 countries in the world. There are about 1,600 accessions of coconut germplasm in the world, which are mainly distributed in Southeast Asia and the South Pacific, where coconut production and hectarage account for about 80% of the total in the world.

The Federated States of Micronesia is known as the Pearl of the Pacific Ocean. There are 607 islands in the territory, and coconuts are the dominant species on the island. We conducted a comprehensive survey of seven major islands in the country, recording 37 different types of coconut germplasm. The botanical morphological characteristics of each accession of germplasm were described in detail, and the production potential and breeding value of each accession of germplasm were initially measured on site. With the limited time, survey area and the insufficient reference materials, the resources described are definitely missing and the information provided is hardly complete. However, this book is the first professional and popular science monograph on the description of coconut germplasm resources in the country. It is of great reference value for the research and planting of coconut in the country.

General Editor

Vice President of Chinese Academy of Tropical Agricultural Sciences

March 22, 2019

Contents

- **Yap State** ················· 1
  - Coconut Yap Female Brown Tall ········· 2
  - Coconut Yap Female Green Tall ········· 3
  - Coconut Yap Triangle Brown Tall ········· 4
  - Coconut Yap Long Triangle Brown Tall   5
  - Coconut Yap Brown Tall ················· 6
  - Coconut Yap Yellow Tall ················· 7
  - Coconut Yap Triangle Yellowish
    Green Tall ················· 8
  - Coconut Yap Long Triangle Yellowish
    Green Tall ················· 9
  - Coconut Yap Green Tall ················· 10
  - Coconut Yap Reddish Brown Dwarf ······ 11
  - Coconut Yap Red Dwarf ················· 12
  - Coconut Yap Yellow Dwarf ················· 13
  - Coconut Yap Green Dwarf················· 14

- **Chuuk State** ················· 15
  - Coconut Chuuk Reddish Brown Dwarf··· 16
  - Coconut Chuuk Grenade Brown Dwarf··· 17
  - Coconut Chuuk Grenade Green Dwarf··· 18
  - Coconut Chuuk Brown Dwarf ················· 19
  - Coconut Chuuk Melon Seed Tall ········· 20

- **Kosrae State** ················· 21
  - Coconut Kosrae Triangle Yellow Tall ··· 22
  - Coconut Kosrae Yellowish Green Tall ··· 23
  - Coconut Kosrae Triangle Green Tall ······ 24
  - Coconut Kosrae Green Tall ················· 25
  - Coconut Kosrae Triangle Brown Tall ··· 26
  - Coconut Kosrae Long Fruit Stipe Tal ··· 27
  - Coconut Kosrae Reddish Brown Dwarf   28
  - Coconut Kosrae Reddish Brown
    Dwarf-Papaya Type ················· 29
  - Coconut Kosrae Red Dwarf ················· 30
  - Coconut Kosrae Brown Tall ················· 31
  - Coconut Kosrae Green Dwarf ················· 32
  - Coconut Kosrae Grenade Green Dwarf··· 33

- **Pohnpei State** ················· 34
  - Coconut Pohnpei Green Dwarf ··········· 35
  - Coconut Pohnpei Green Tall ················· 36
  - Coconut Pohnpei Brown Dwarf············ 37
  - Coconut Pohnpei Yellow Dwarf ········· 38
  - Coconut Pohnpei Red Dwarf················· 39
  - Coconut Pohnpei Brown Tall················· 40
  - Coconut Pohnpei Triangle Brown Tall ··· 41

Yap State[1]

Yap is one of the four states of the Federated States of Micronesia (FSM). It consists of Yap islands and the island of Satawal as well as 14 atolls such as Eauripik, Elato, Fais, Faraulep, Gaferut, Ifalik, Lamotrek, Ngulu, Olimarao, Piagailoe (West Fayu), Pikelot, Sorol, Ulithi and Woleai. Administratively, the entire state is home to about 10,000 residents, consisting of 10 villages (also municipalities) such as Dalipebinaw, Fanif, Gagil, Gilman, and Kanifay.

Flag of Yap State

Yap is known for its stone money "Rai" or "Fei", one of the original currencies in ancient times. Since metal is not produced locally, the stone becomes an important resource in the local area and develops into a trade model in which stone acts as a medium for exchange in trading. And it is still used today. The locals call this stone "Rai" or "Fei". Many of them are brought from other islands in New Guinea, but most of them come from ancient Palau.

The largest local stone coin

Women dancing in folk costume

Bags made with palm leaves

Traditional men's house

---

[1] https://en.wikipedia.org/w/index.php?title=Yap&oldid=1026401357

Coconut Yap Female Brown Tall

English name: Yap Female Brown Tall (YFBT)

Main features:

YFBT is a tall coconut type. The palm is tall and stout. The trunk diameter is 90–120 cm, the height can be more than 20 meters, and the trunk base is swollen to form a shape of bottle gourd. The crown is round and consists of 25–30 fronds and each frond is 400–500 cm long. It can bear fruits after 7–8 years of planting. Its economic life can be 60–80 years long and the natural life span is more than 100 years.

The palm has few male flowers with 1–2 male inflorescences and more female flowers which are grouped in a spindle shape. It is also known as the female palm. Some palms have incompletely pinnatified fronds on the top with 4–5 leaflets bound together. The leaflets are wider, about 7 cm wide and the inflorescence is short, only 70–90 cm long. It is often difficult to set fruits due to poor pollination, and the palm yield low due to abnormal development of the fruits. The fruit is elliptical and brown; the shell is round, but often oblate. The flesh is medium thick.

This variety is only distributed in a small area, and grows mostly in single occasionally, 2–3 palms grow together and are often mistaken as male coconut palms. They are found in the main islands of Yap and the outer island, and can be collected and conserved as a genetic resource. In the local area, this variety is not used and mostly abandoned.

Coconut Yap Female Green Tall

English name: Yap Female Green Tall (YFGT)

Main features:

YFGT is a tall coconut variety. The palm is tall and stout, trunk diameter is about 90–120 cm and the height is more than 20 meters. The stem base is swollen which is called "bottle gourd". The crown is round and consists of 25–30 fronds which are 400–500 cm long. It can bear fruits after 7–8 years of planting. Its economic life can be 60–80 years long and the natural life span is more than 100 years.

The palm has few male flowers with 1–2 male inflorescences and more female flowers, which are gathered into a spindle shape. It is also known as the female tree, characterized with incompletely pinnatified pinnae on the top in some palms with 4–5 leaflets bound together. The leaflet is wider, and it is about 7 cm wide. The inflorescence is a bit short, only 70–90 cm. The palm is often difficult to set fruits due to poor pollination and yields low due to abnormal development of the fruits. The fruit is green, elliptical, thick in flesh, and the shell is round and often oblate.

This variety is only distributed in a small area and found to grow in single. Occasionally, 2–3 palms are found to grow together and often mistaken as female palms. They are only found on the outer island and can be collected and conserved as a genetic resource. In the local area they are not used and mostly abandoned.

Coconut Yap Triangle Brown Tall

English name: Yap Triangle Brown Tall (YTBT)

Main features:

YTBT belongs to a tall-coconut type. This palm is tall and thick with trunk diameter of 90–120 cm. The palm can be more than 20 meters tall, and the trunk base is swollen to form a shape of bottle gourd. The crown is round, consisting of 30–40 fronds. The fronds are about 500–600 cm long. It can set fruits after 7–8 years of planting. Its economic life can be 60–80 years long, and the natural life span is more than 100 years.

YTBT is monoecious and produces many fruits with an annual average yield of 100 fruits, even more than 150 fruits. The coconut size is medium. The fruit is 3 angledat the top and the flesh is thick. The shell is round and its color is brown.

This variety is widely distributed in the main island Japu and the outer islands. It is a conventional cultivar from which local residents mostly take the flesh for food. This palm should be a locally domesticated elite variety.

Coconut Yap Long Triangle Brown Tall

English name: Yap Long Triangle Brown Tall (YLTBT)

Main features:

YLTBT belongs to a tall- coconut type. The palm is tall and thick with the trunk diameter of 90–120 cm. The palm is more than 20 meters tall and has a swollen trunk base with a shape of gourd head. The crown is round, consisting of 30–40 fronds which are 500–600 cm long. The palm can set fruits 7–8 years after planting. Its economic life can be 60–80 years long and the natural life span is more than 100 years.

YLTBT is monoecious and yields moderate with an annual average yield of about 80 fruits. The coconut is long triangular at the top. The flesh is thick and the shell is round. The outer husk is brown.

Coconut Yap Brown Tall

English name: Yap Brown Tall (YBT)

Main features:

YBT is a tall coconut variety. The palm is tall and stout with the trunk diameter of 90–120 cm. The palm can be more than 20 meters tall and has a swollen trunk base with a shape of bottle gourd. The crown is round, consisting of 30–40 fronds which are 500–600 cm long. The palm can bear fruits 7–8 years after planting. The economic life can be 60–80 years long and the natural life span is more than 100 years.

YBT is monoecious. The fruit yield is moderate, 90–100 fruits/year. The size of coconut fruit is medium, smaller than that of the coconut Triangle Brown Tall. The fruit is round and the flesh is thick. The shell is round. The outer husk is brown.

This variety is widely distributed in the main island Japu and the outer islands. It is also one of the conventional cultivars for the local residents. It is commonly found on both sides of the roads and around the villages.

Coconut Yap Yellow Tall

English name: Yap Yellow Tall (YYT)

Main features:

YYT is a tall-coconut variety. The palm is tall and stout with the trunk diameter of 90–120 cm. The palm can be more than 20 meters tall, and has a swollen trunk base which is called "bottle gourd". The crown is round, consisting of 30–40 leaves. The leaves are 500–600 cm long.

YYT is monoecious with cross-pollination. The annual average yield is about 100 fruits. The coconut fruit size is medium. The fruit is yellow and round, and the flesh is thick. The shell is round.

It is currently only found at the entrance of the Main Agricultural Training Center in the Yap Island. This variety has not been found on the outer islands. It should be a mutant which may have high research value in breeding.

Coconut Yap Triangle Yellowish Green Tall

English name: Yap Triangle Yellowish Green Tall (YTYGT)

Main features:

YTYGT is a tall variety. The palm is tall and stout with the trunk diameter of 90–120 cm. The height can be more than 20 meters, and the trunk base is swollen to have a shape of gourd head. The crown is round, consisting of 30–40 fronds which are 500–600 cm long. This variety can set fruits 7–8 years after planting.

YTBT is monoecious and is cross-pollinating. The yield is moderate, about 80 fruits/year. The coconut size is medium. The fruit is yellowish green, the apex of the fruit is often triangular and the flesh is thick. The shell is round.

This variety is widely distributed in the main island and the outer islands of Yap. It is one of the conventional cultivars for the local residents, and it is also the source of local edible coconut. It should be the local dominant domesticated variety as the coconut Triangle Brown Tall.

Coconut Yap Long Triangle Yellowish Green Tall

English name: Yap Long Triangle Green Tall(YLTGT)

Main features:

YLTGT is a tall-coconut type. The palm is tall and thick with the trunk diameter of 90–120 cm. The palm is more than 20 meters tall, and its trunk base is swollen to form a shape of gourd head. The crown is round, semi-circular, and Y-shaped, consisting of 30–40 fronds which are 380–450 cm long. It can set fruits after 7–8 years of planting. The economic life span can be 60–80 years long, and the natural life span is more than 100 years.

All palms of this variety are heterozygous. The coconut is small and the copra is of high quality. In the frond 4–5 leaflets are bonded together. The leaflets are about 7 cm wide (normally 4–5 cm), and the frond is short, about 400 cm long (normally 450–600 cm). The inflorescence is short, about 70–90 cm long (normally 110–120 cm). This variety is monoecious. The male and female phases often overlap each other in the same inflorescence, and the palm is self-pollinating. The palm produces numerous fruits, with an annual average yield of more than 100 fruits, even more than 150 fruits. The fruit is triangular, green and long in shape and the flesh is thick. The shell is round and often oblate.

Coconut Yap Green Tall

English name: Yap Green Tall (YGT)

Main features:

YGT is a tall- cocoanut variety. The palm is tall and thick with the trunk diameter of 90–120 cm. The palm can grow as tall as more than 20 meters, and the trunk base is swollen with a shape bottle gourd. The crown is round, consisting of 30–40 fronds which are 500–600 cm long. This variety can set fruits after 7–8 years of planting.

YTBT is monoecious, and produces many fruits with an annual average yield of about 80 fruits. The coconut size is medium. The fruit are mostly round or triangular at the top and the flesh is thick. The shell is round and the outer husk is green.

This variety is distributed on the main island of Yap, and occasionally on the outer island.

Coconut Yap Reddish Brown Dwarf

English name: Yap Reddish Brown Dwarf (YRBD)

Main features:

YRBD is a typical dwarf coconut. The palm is short with the trunk diameter of 70–80 cm. The trunk base is not swollen with no shape of bottle gourd or a shape of small bottle gourd. The crown is round, consisting of 20–30 fronds which are 450–550 cm long. This variety can produce fruits after 3–4 years of planting.

YRBD is self-pollinating and produces many fruits with an average annual yield of more than 100 fruits. The fruit is ovate and the flesh is thinner than that of the tall type. The husk is thinner than that of the common tall type. The shell is round, often oblate. The sugar content of the coconut water is higher than that of the tall variety, suitable for fresh consumption. The color of the outer husk is reddish brown, due to this variety is suitable for landscaping or planting at courtyards.

Coconut Yap Red Dwarf

English name: Yap Red Dwarf (YRD)

Main features:

YRD is a typical dwarf type. The palm is short and the trunk diameter is 70–80 cm. The trunk base is not swollen and has no shape of bottle gourd or a shape of small bottle gourd. The crown is round, consisting of 20–30 fronds which are 450–550 cm long. This variety can produce fruits after 3–4 years of planting.

YRD is self-pollinating and produces many fruits with an average annual yield of more than 100 fruits. The fruit is ovate and the coconut flesh is thinner than that of the tall type. The husk is thinner than that of the common tall coconut. The shell is round and often oblate. The sugar content of coconut water is higher than that of the tall type, and the coconut fruits are suitable for fresh consumption. The outer husk is orange-red in color. This variety is often planted for landscaping or at courtyards for greening due to its fruit color. It is distributed in Yap and is often planted for greening at the courtyard.

Coconut Yap Yellow Dwarf

English name: Yap Yellow Dwarf (YYD)

Main features:

YYD is a typical dwarf type. The palm is short and the trunk diameter is 70–80 cm. The trunk base is not swollen and has no shape of bottle gourd or a shape of small bottle gourd. The crown is round, consisting of 20–30 fronds which are 450–550 cm long. It can bear fruits after 3–4 years of planting.

YYD often has an overlap of its male and female phases in the same influorescence and is self-pollinating. It produces more fruits with an annual average yield of more than 100 fruits. The fruit is round and golden yellow. The coconut flesh is thinner than that of the tall type. The husk is thinner and the shell is round, often oblate. The coconut water has higher sugar content than that of the other tall varieties. The coconut flesh is soft, suitable for fresh consumption.

This variety is mainly distributed on the main island of Yap, and is often planted in gardens and parks or courtyards for greening.

Coconut Yap Green Dwarf

English name: Yap Green Dwarf (YGD)

Main features:

YGD is a typical dwarf type. The palm is short with the trunk diameter of 70–80 cm. The trunk base is not swollen and does not have a shape of bottle gourd or have a shope of small bottle gourd. The crown is round, consisting of 20–30 fronds. Each frond is 450–550 cm long. This variety can produce fruits after 3–4 years of planting.

YGD often has an overlap of both its male and female phases in the same inflorescence and is self-pollinating. It produces many fruits with an average annual yield of more than 100 fruits. The fruit is round. The coconut flesh is thinner than that of the tall growing type. The coconut husk is thinner than that of the tall varieties. The shell is round and often oblate. The coconut water has higher sugar content than the other tall varieties. The coconut flesh is soft, suitable for fresh consumption. The outer husk is emerald green.

This variety can be found on the main island of Yap and the outer island. It is often used for fresh consumption or making wine by using coconut inflorescence sap in the outer island.

Chuuk State[1]

Chuuk is one of the four states of the Federated States of Micronesia (FSM). It consists of several island groups: Chuuk Lagoon, Nomwisofo, Hall Islands, Namonuito Atoll (Margull Islands), Pattiw (Western Islands), Eastern Islands (Upper Lockheed Islands) and Mortlock .

Chuuk flag

Chuuk is the most populous state in FSM, with 500 square kilometers and a population of 54,595. Chuuk Lagoon is the place where most people live. The Weno Island in the Lagoon is the capital of the state and is the largest city in FSM.

Chuuk women's traditional dress

Traditional celebration traditional dress

[1] https://en.wikipedia.org/w/index.php?title=Chuuk_State&oldid=1023106815

Coconut Chuuk Reddish Brown Dwarf

English name: Chuuk Reddish Brown Dwarf (CRBD)

Main features:

CRBD is a typical dwarf coconut. The palm is short and the trunk diameter is 70–80 cm. The trunk base is not swollen and has no shape of bottle gourd or a shape of small bottle gourd. The crown is round, consisting of 20–30 fronds which are 450–550 cm long. It can produce fruits after 3–4 years of planting.

CRBD is self-pollinating and produces many fruits. The average annual yield is more than 120 fruits, even more than 300 fruits. The fruit is elliptical. The coconut water is sweet and the coconut flesh is soft. The mature coconut flesh is thinner than that of the other tall varieties. The shell is round and often oblate. The outer husk is bright reddish brown.

This variety is the most distributed on the Weno Island of the main island of Chuuk State. It is mostly used for greening in gardens and parks and fresh consumptionfruit. This variety is also often found along the sides of the roads.

Coconut Chuuk Grenade Brown Dwarf

English name: Chuuk Grenade Brown Dwarf (CGBD)

Main features:

CGBD is the typical dwarf coconut. The palm is short and the trunk diameter is 70–80 cm. The trunk base is not swollen and has a shape of small bottle gourd. The crown is round, consisting of 20–30 fronds. Each frond is 450–550 cm long. It can produce fruits after 3–4 years of planting.

CGBD has an overlap of its male and female phases in the same influorescence and is self-pollinating. It produces many fruits with an annual average yield of more than 150 fruits, even more than 200 fruits. The fruit is small, one third of that of the common tall type; the fruit is oval, similar to a shape of a grenade. The flesh is thinner than that of the other tall and common dwarf varieties and the shell is round and often oblate. The outer husk is brown in color. This variety is widely distributed in Weno Island of chuuk. It should be a coconut variety artificially selected and is often used for fresh consumption.

Coconut Chuuk Grenade Green Dwarf

English name: Chuuk Grenade Green Dwarf (CGGD)

Main features:

CGGD is a typical dwarf coconut. The palm is short with the trunk diameter of 70–80 cm. The trunk base is not swollen and has a shape of small bottle gourd. The crown is round, consisting of 20–30 fronds which are 450–550 cm long. It can set fruits after 3–4 years of planting.

CGGD has the same flowering stage for the male and female flowers and it is self-pollinating. This variety produces many fruits with the annual average yield of more than 150 fruits, even more than 300 fruits. The fruit is small and its size is only one third of that of the common tall. The fruit is oval and similar to a shape of a grenade. The flesh is thinner than that of the tall and common dwarf varieties. The shell is round and often oblate. The outer husk is emerald green. This variety is widely distributed in the Weno Island of Chuuk State. It should be a cultivar artificially selected and often used for fresh consumption.

The fruit has obvious characteristic of high-yield, excellent quality, and sweet coconut water, and is hence of high breeding value.

Coconut Chuuk Brown Dwarf

English name: Chuuk Brown Dwarf (CBD)

Main features:

The CBD is the typical dwarf coconut. The palm is short with the trunk diameter of 70–80 cm. The trunk base is not swollen and has a shape of small bottle gourd. The crown is round, consisting of 20–30 fronds. Each frond is 450–550 cm long. It can produce fruits after 3–4 years of planting.

CBD has the same flowering stage for the male and female flowers and is self-pollinating. This variety produces more than 100 fruits per year. The fruit is small and oval with a convex top. The coconut flesh is thinner than that of the tall and other common dwarf varieties. The shell is ovate, but often oblate. The outer husk is yellowish brown.

The variety has obvious characteristics of high yield and high fruit quality with sweet coconut water, and has certain value for breeding.

Coconut Chuuk Melon Seed Tall

English name: Chuuk Melon seed Tall (CMST)

Main features:

CMST is the tall coconut type. The palm is tall and thick with the trunk diameter of 90–120 cm. The tree height can be as tall as more than 20 meters, and the trunk base is swollen and has a shape of bottle gourd. The crown is round, consisting of approximately 30 fronds. Each frond is 500–600 cm long. This variety can produce fruits after 7–8 years of planting.

CMST is monoecious. The fruit is very large with an annual average yield of about 160 fruits. The fruit is reddish brown and slender, and its top is sharp, similar to the shape of a melon seed. The husk of mature fruit accounts for three quarters of the fruit with a small seed. The shell is similar to that of a melon seed. The inner side of the shell is waterless and the flesh is thick.

CMST should be a mutant from a variety with a small distribution. Only two palms were found near the University of Crick, Chuuk. They are not used by local residents but can be used as a genetic resource or for landscaping.

● Kosrae State[①]

Kosrae is one of the four states of the Federated States of Micronesia (FSM) and Tofol is the state capital. Kosrae has only the island and no outer island (The island has been integrated with its five small islands) with a land area of 110 square kilometers and a population of 6,600.

There are three main topographic features on the island: mountains, rainforests and mangroves. The mountainous area of the island accounts for 70% of the total area of the island and the highest peak is over 2000 meters. If you look at the island from the sea, the outline of the island is very similar to a lying woman. As a result, it is also known as the "Island of Sleeping Beauty". Most parts of the island have not yet been developed.

Kosrae flag

Diving spot—Kosrae

---

① https://en.wikipedia.org/w/index.php?title=Kosrae&oldid=1015000744

Coconut Kosrae Triangle Yellow Tall

English name: Kosrae Triangle Yellow Tall (KTYT)

Main features:

KTYT is a tall coconut variety. The palm is tall and stout with the trunk diameter of 90–120 cm. The palm can be more than 20 meters tall and its trunk base is swollen to form a gourd head shape. The crown is round, consisting of 30–40 fronds. The frond length is 500–600 cm. The variety can produce fruits 6 years after planting and its life span can be more than 40 years.

KTYT is monoecious, and produces many fruits with an annual average yield of about 100 fruits. The fruit is large, elliptical with a triangular shape at the top. The size is close to that of the Kosrae Triangle Brown Tall, and is four times as big as that of the common yellow dwarf coconut. The shell is round or oval. The outer husk is golden yellow.

This variety should be a mutant from a yellow dwarf or a brown tall type. It is also possible to be a hybrid between the brown tall and the yellow dwarf varieties. The variety is less distributed and only two palms were found in Tofol, Kosrae.

Coconut Kosrae Yellowish Green Tall

English name: Kosrae Yellowish Green Tall (KYGT)

Main features:

KYGT is a typical tall coconut. The palm is tall and stout with a trunk diameter of 90–120 cm, and the height can be more than 20 m. The trunk base is swollen to form a shape of bottle gourd. The crown is round, consisting of 30–40 fronds. Each frond is 500–600 cm long. This variety can produce fruits after 7–8 years of planting.

KYGT is monoecious and produces less fruits. The coconut fruit is large and oval. The size is close to that of the Triangle Brown Tall. The shell is round or oval. The outer husk is yellowish green.

The variety is widely distributed in Kosrae, and can be seen along the coastal roads and around the village.

Coconut Kosrae Triangle Green Tall

English name: Kosrae Triangle Green Tall (KTGT)

Main features:

KTGT is a typical tall coconut type. The palm is tall and stout with the trunk diameter of 90–120 cm. The palm can grow more than 20 meters tall, and the trunk base is swollen to form a shape of bottle gourd. The crown is round, consisting of 30–40 fronds. Each frond is 500–600 cm long. This variety can produce fruits after 7–8 years of planting.

KTGT is monoecious, and produces many fruits with an annual average yield of about 100 fruits. The coconut fruit is large, ovate with a triangular shape at the top. The fruit size is close to that of the Kosrae Triangular Brown Tall coconut variety. The shell is round or oval and the outer husk is green.

The variety is widely distributed in Kosrae and can be seen along the coastal roads and around the village.

Coconut Kosrae Green Tall

English name: Kosrae Green Tall (KGT)

Main features:

KGT is a typical tall coconut type. The palm is tall and stout with a trunk diameter of 90–120 cm. The palm height is more than 20 m. The trunk base is swollen to form a shape of gourd head. The crown is round, consisting of 30–40 fronds. Each frond is 500–600 cm long. This variety can produce fruits after 7–8 years of planting.

KGT is monoecious, and produces many fruits with an annual average yield of about 100 fruits. The coconut fruit is oval. The size is medium. The shell is round or oval, and the outer husk is green.

The variety is widely distributed in Kosrae and can be seen along the coastal roads and around the village.

Coconut Kosrae Triangle Brown Tall

English name: Kosrae Triangle Brown Tall (KTBT)

Main features:

KTBT is a typical tall coconut type. The palm is tall and stout, with a trunk diameter of 90–120 cm, the height can be more than 20 m, and the trunk base is swollen to form a shape of bottle gourd. The crown is round, consisting of 30–40 fronds. Each frond is 500–600 cm long. This variety can produce fruits after 7–8 years of planting.

KTBT is monoecious, and produces an annual average yield of about 100 fruits. The coconut fruit is large, oval, triangular at the top. Its size is close to that of Triangle Brown Tall. The shell is round or oval. The color of the outer husk is yellowish brown.

The variety is widely distributed in Kosrae and can be seen along the coastal roads and around the village.

Coconut Kosrae Long Fruit Stipe Tall

English name: Kosrae Long Fruit Stipe Tall (KLFST)

Main features:

KLFST is a typical tall coconut type. The palm is tall and stout, with a trunk diameter of 90–120 cm, and the height can be more than 20 m, and the trunk base is swollen to form a shape of bottle gourd. The crown is round, consisting of 30–40 fronds. Each frond is 500–600 cm long. This variety can produce fruits after 7–8 years of planting.

KLFST is monoecious, and produces many fruits. The annual average yield is about 100 fruits. The coconut is large and its size is close to that of the Triangle Brown Tall coconut variety. The shell is round or oval. The outer husk is yellowish brown.

The variety has an inflorescence with a long peduncle and is an important genetic resource. A small distribution of KLFST can be found around the Kosrae coastal roads and around the village.

Coconut Kosrae Reddish Brown Dwarf

English name: Kosrae Reddish Brown Dwarf (KRBD)

Main features:

KRBD is a typical dwarf coconut. The palm is short, and its trunk diameter is 70–80 cm. The trunk base is not enlarged, with a shape of small bottle gourd. The crown is round, consisting of 20–30 fronds. Each frond is 450–550 cm long. This variety can produce fruits after 3–4 years of planting.

KRBD often has the same flowering stage for male and female flowers in the same inflorescence and is self-pollinating. It produces many fruits, with an annual average yield of more than 200 fruits. The fruit is small and oval. Its flesh is thinner than that of the tall and common dwarf types. The shell is round and often oblate. The outer husk is yellowish brown.

The fruit has obvious high yield and excellent quality with sweeter coconut water, and this variety has certain value for breeding. This variety is cultivated by local residents and is often used for fresh consumption.

Coconut Kosrae Reddish Brown Dwarf-Papaya Type

English name: Kosrae Reddish Brown Dwarf- Papaya type (KRBD-P)

Main features:

KRBD-P is a typical dwarf coconut. The palm is short, the trunk diameter 70–80cm, and the trunk base is not enlarged, with a shape of small bottle gourd. The crown is round, consisting of 20–30 fronds. Each frond is 500–600 cm long. This variety can produce fruits after 3–4 years of planting.

KRBD-P is monoecious. The male and female phases often overlap each other in the same inflorescence, and and this variety is self-pollinating. The variety produces many fruits with an annual average yield of more than 120 fruits, even more than 150 fruits. The fruit is long elliptical, which resembles papaya. The shell is round, often oblate. The outer husk is reddish brown.

Coconut Kosrae Red Dwarf

English name: Kosrae Red Dwarf (KRD)

Main features:

KRD is a typical dwarf coconut. The palm is short, the trunk diameter 70–80 cm, the trunk base is not enlarged, with a shape of a small bottle gourd. The crown is round, consisting of 20–30 fronds. Each frond is 400–500 cm long. This variety can produce fruits after 3–4 years of planting.

KRD is monoecious, and its male and female flowers often occur at the same flowering stage. The variety is self-pollinating. The fruit is round or elliptical. The flesh is thinner than that of the tall coconut type. The shell is round or oval. The outer husk is orange.

The variety is less distributed in Kosrae and is found in the village or courtyards.

Coconut Kosrae Brown Tall

English name: Kosrae Brown Tall (KBT)

Main features:

The KBT palms are robust, medium height with an average trunk circumference of 90–120 cm. The trunk base is swollen to form a distinct shape of bottle gourd. The round crown has 35–45 fronds. The fronds are long with a strong petiole. It can produce fruits after 7 years of planting. The inflorescence is long with thick pedicels. It is monoecious and self-pollinating due to the fact that male and female phases often overlap each other in the same inflorescence. The variety produces fruits 7–8 years after planting with 8 to 10 bunches per palm per year. And 70–90 fruits can be produced each year.

The fruit is large and round, and brown in color. The shell is nearly round and thick. The base of the fruit is flat and the flesh is thick.

Coconut Kosrae Green Dwarf

English name: Kosrae Green Dwarf (KGD)

Main features:

KGD is a typical dwarf coconut. The palm is short with the trunk diameter of 70–80 cm. The trunk base is not swollen and has no shape of bottle gourd or a shape of small bottle gourd. The crown is round and semi-circular, consisting of 20–30 fronds. Each frond is 500–600 cm long. This variety can produce fruits after 3–4 years of planting.

KGD is monoecious The male and female phases often overlap each other in the same inflorescence. This variety is self-pollinating and produces many fruits with an annual average yield of more than 100 fruits, even more than 150 fruits. The fruit is round and elliptical. The flesh is thinner than that of the tall coconut type. The shell is round, often oblate. The outer husk is green.

The variety is morphologically close to the Marshall Green Dwarf coconut, and is cultivated locally for fresh consumption. It is cultivated in the periphery of the courtyards, and the local farmers often propagate this variety for cultivation by themselves because of the good quality of the fruit.

Coconut Kosrae Grenade Green Dwarf

English name: Kosrae Grenade Green Dwarf (KGGD)

Main features:

KGGD belongs to the dwarf coconut type. The palm is short with the trunk diameter of 70–80 cm. The trunk base is not swollen and has no shape of bottle gourd or a shape of small bottle gourd. The crown is round and semi-circular, consisting of 20–30 fronds. Each frond is 400–500 cm long. It can produce fruits after 3–4 years of planting.

The number of female flowers is large, and there are 5–6 female flowers in each spikelet. The female flowers are numerous and the fruit drops are very serious due to poor pollination or fertilizer competition. This variety produces many fruits with an annual average yield of more than 100 fruits, even more than 150 fruits. The fruit shape like a grenade, the flesh is thinner than that of the tall type. The shell is round, often oblate. The outer husk is green.

● Pohnpei State[①]

Pohnpei is one of the four states of the Federated States of Micronesia (FSM). It is made up of Ponape Island and its seven surrounding atolls and is part of the Caroline Islands. Palikir, the FSM capital is located in the state. The capital of Pohnpei is Colonia in the north of Pohnpei.

Pohnpei is the largest, highest-lying and most populous atoll island in the Federated States of Micronesia. It has a population of about 34,000 and is dominated by the Pohnpeian. The highest elevation of Pohnpei is 760 m, which is also the highest point of Pohnpei. Pohnpei also includes Pingelap Island, which has a very high proportion of completely color blind populations and a unique Pingelapese language.

Pohnpei flag

Women's traditional costumes in Pohnpei

Making traditional drinks Sakau

C-K Friendly Sports Center

---
① https://en.wikipedia.org/w/index.php?title=Pohnpei_State&oldid=1023985909

Coconut Pohnpei Green Dwarf

English name: Pohnpei Green Dwarf (PGD)

Main features:

PGD is a typical dwarf coconut. The palm is short with a trunk diameter of 70–80 cm. The trunk base is not swollen and has no shape of bottle gourd or a shape of small bottle gourd. The crown is round or semi-circular, consisting of 30–40 fronds. Each frond is 500–600 cm long. It can produce fruits after 3–4 years of planting. The economic life span is 60–80 years and the natural life span is more than 100 years.

PGD often has an overlap of its male and female phases in the same inflorescence, and is self-pollinating. It produces many fruits with an annual average yield of more than 100 fruits, even more than 150 fruits. The fruit is round with sweet coconut water and fluffy coconut flesh. The mature coconut flesh is thinner than that of the tall type. The husk is thick and the shell is smaller, round and often oblate. The outer husk is green.

This variety is widely distributed in Pohnpei Island and is one of the conventional cultivars favored by the local population. It can be commonly found on both sides of the roads and around the villages.

Coconut Pohnpei Green Tall

English name: Pohnpei Green Tall (PGT)

Main features:

PGT is a tall coconut type. The palm is tall and stout with the trunk diameter of 90–120 cm. The height can be more than 20 meters and the trunk base is expanded to form a shape of bottle gourd. The crown is round, consisting of 30–40 fronds. Each frond is 500–600 cm. This variety can produce fruits about 6 years after planting, and its life span can be more than 40 years.

The variety is monoecious and cross-pollinating. It produces many fruit with an annual average yield of about 120 fruits. The fruit is large and elliptical, with a triangular shape at the top. The shell is round or oval. The outer husk is green.

This variety is widely distributed in Pohnpei Island and is one of the conventional cultivars for the local people. It can be commonly found on both sides of the road and around the village.

Coconut Pohnpei Brown Dwarf

English name: Pohnpei Brown Dwarf (PBD)

Main features:

PBD is a typical dwarf coconut. The palm is short with the trunk diameter of 70–80 cm. The trunk base is not swollen and has no shape of bottle gourd or a shape of small bottle gourd. The crown is round and semi-circular, consisting of 30–40 fronds. Each frond is 500–600 cm long. It can bear fruits after 3–4 years of planting. The economic life span can be 60–80 years and the natural life span is more than 100 years.

PBD often overlaps in its male and female phases in the same inflorescence and is self-pollinating. It produces many fruits with an annual average yield of more than 100 fruits, even more than 150 fruits. The fruit is round with a triangular shape at the top. The coconut water is sweet and the flesh is soft. The mature coconut flesh is thinner than that of the tall type. The shell is round and often oblate. The outer husk is brownish yellow.

Coconut Pohnpei Yellow Dwarf

English name: Pohnpei Yellow Dwarf (PYD)

Main features:

PYD is a typical dwarf coconut. The palm is short with the trunk diameter of 70–80 cm. The trunk base is not enlarged and has no shape of bottle gourd or a shape of small bottle gourd. The crown is round and semi-circular, consisting of 26–38 fronds. Each frond is 420–580 cm long. It can bear fruits after 3–4 years of planting. The economic life span can be 60–80 years and the natural life span is more than 100 years.

PYD often overlaps in its male and female phases in the same inflorescence and is self-pollinating. It produces many fruits with an annual average yield of more than 100 fruits, even more than 150 fruits. The fruit is round and the coconut water is sweet. The coconut flesh is soft and the mature flesh is thinner than that of the tall type. The shell is round and often oblate. The outer husk is yellow.

Coconut Pohnpei Red Dwarf

English name: Pohnpei Red Dwarf (PRD)

Main features:

PRD is a typical dwarf coconut. The palm is short with the trunk diameter of 70–80 cm. The trunk base is not enlarged and has no shape of bottle gourd or a shape of small bottle gourd. The crown is round and semi-circular, consisting of 30–40 fronds. Each frond is 430–520 cm long. It can bear fruits after 3–4 years of planting. The economic life span can be 40–50 years and the natural life span is over 80 years.

PRD often overlaps in the male and female phases in the same inflorescence and is self-pollinating. It produces many fruits with an annual average yield of more than 120 fruits, even more than 200 fruits. The fruit is round or elliptical, and the flesh is thinner than that of the tall type. The shell is round, often oblate. The outer husk is bright orange-red, more vivid than that the Yap Red Dwarf. This variety is suitable for fresh consumption as well as for landscaping or courtyard planting.

Coconut Pohnpei Brown Tall

English name: Pohnpei Brown Tall (PBT)

Main features:

PBT is a tall type. The palm is tall and stout with the trunk diameter of 90–120 cm. The height can be more than 20 meters. The trunk base is expanded to form a shape of bottle gourd. The crown is round, consisting of 30–40 fronds. Each frond is 500–600 cm long. It can bear fruits after 7–8 years of planting. The economic life span is 60–80 years and the natural life span is more than 100 years.

PBT is monoecious, and yields moderate with an annual average yield of 90–100 fruits. The coconut fruit is medium sized, smaller than the Triangle Brown Tall. The fruit is round and the flesh is thick. The shell is round. The outer husk is brown.

The variety is widely distributed in Pohnpei Island and is one of the conventional cultivars for the local people. It is commonly found on both sides of the road and around the village.

Coconut Pohnpei Triangle Brown Tall

English name: Pohnpei Triangle Brown Tall (PTBT)

Main features:

PTBT is a tall coconut type. The palm is tall and stout with the trunk diameter of 90–120 cm. The height can be more than 20 meters. The trunk base is expanded to form a shape of bottle gourd. The crown is round, consisting of 28–36 fronds. Each frond is 480–570 cm long. It can bear fruits after 7–8 years of planting. The economic life span can be 60–80 years and the natural life span is over 100 years.

PTBT is monoecious, and bears many fruits with an annual average yield of 120 fruits, even more than 150 fruits. The coconut is medium sized. The top of the fruit is triangular and the flesh is thick. The shell is round. The outer husk is brown.

This variety is distributed in the island of Pohnpei and is a conventional cultivar for local residents.

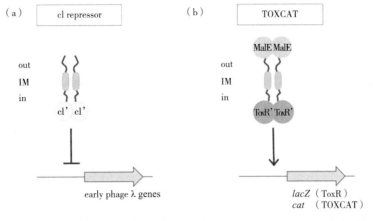

图16-1 用于监测TMH同型相互作用的示意图

（a）cI阻遏蛋白检测。目标TMH（橙色）与cI DNA结构域（绿色）融合。重组cI二聚体抑制λ噬菌体早期基因转录表达（蓝色）。λ噬菌体基因的抑制赋予对λ噬菌体感染的保护。（b）ToxR和TOXCAT测定。目标TMH（橙色）融合在霍乱弧菌ToxR DNA结构域（红色）和MalE蛋白（蓝色）之间。重组ToxR二聚体诱导报告基因的转录表达（蓝色，用ToxR测定lacZ，用TOXCAT测定cat）。

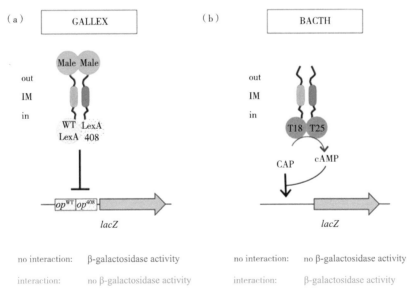

图16-2 用于监测TMH异型相互作用的示意图

（a）GALLEX检测。第1个目标TMH（橙色）融合在野生型LexA DNA结构域（WT LexA）和MalE之间，而第2个TMH（蓝色）融合在LexA408突变体（LexA408）和MalE之间。LexAWT/LexA408二聚体的重构抑制报告基因的转录表达（蓝色）。（b）BACTH检测。第1个目标TMH（橙色）与百日咳博德特氏菌腺苷酸环化酶的T18结构域融合，而第2个TMH（蓝色）与腺苷酸环化酶的T25结构域融合。重组体T18/T25腺苷酸环化酶导致产生环腺苷一磷酸（cAMP）。cAMP与分解代谢物激活蛋白（CAP）的结合诱导报告基因的转录表达（蓝色）。

（宫晓炜 译）

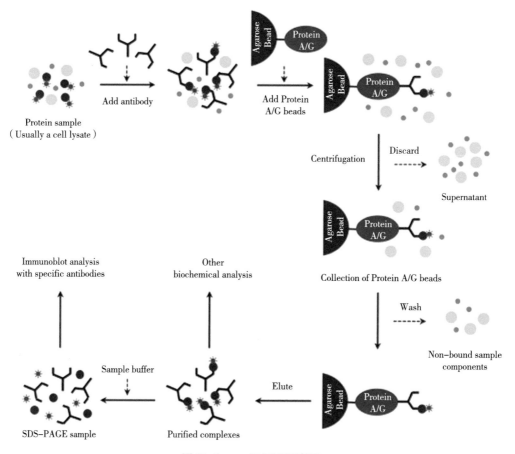

图17-1 co-IP原理示意图

依次加入含抗原的蛋白质样品（通常是细胞裂解物）、特异性抗体和亲和珠（通常是蛋白A/G，其可以特异性结合抗体的保守区域）用于结合反应。通过离心收集具有结合蛋白的亲和珠。弃去含有未结合蛋白质的上清液，并在洗涤步骤中进一步被除去。用缓冲液洗脱抗体和抗原，将蛋白质与亲和珠分离。纯化的蛋白质复合物可进一步用于免疫印迹或其他生化分析。

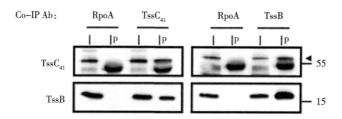

图17-2 根瘤农杆菌中TssB和$TssC_{41}$的Co-IP分析

用DTBP交联剂处理的根瘤农杆菌野生型菌株C58分离的总蛋白质提取物进行1%SDS的缓冲液溶解，然后在含有Triton X-100的IP溶液中稀释。通过蛋白质印迹鉴定共沉淀的蛋白。使用针对RNA聚合酶α亚基（RpoA）的抗体作为阴性对照进行Co-IP实验。分析蛋白质的名称和分子量标准的大小分别在左侧和右侧显示，必要时用箭头表示（I免疫印迹、IP免疫沉淀）[转载自参考文献[10]，重复使用公共科学图书馆（PLOS）发布的内容不需要任何许可]。

（宫晓炜 译）

附 图

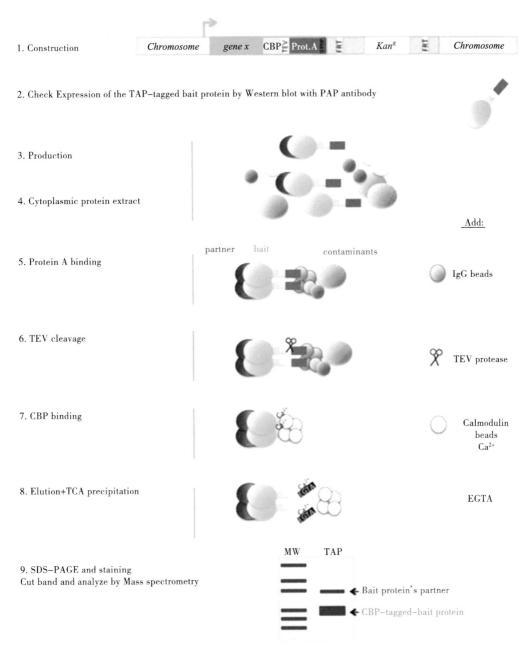

图18-1 整体TAP方法指南

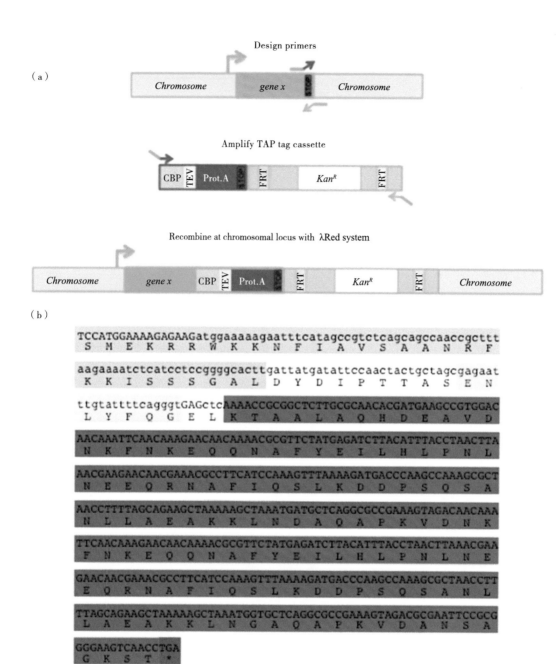

图18-2　在染色体基因座处C-末端T引入TAP-标签融合体

染色体基因座上的TAP-标签融合方案（a）相应的核苷酸和蛋白质TAP-标签序列（b）。CBP序列为浅紫色；核苷酸CBP序列开头的大写字母对应于本章3.1小节中提到的引物序列，TEV蛋白酶切割位点为黄色，ProtA序列为深紫色，终止密码子为红色。

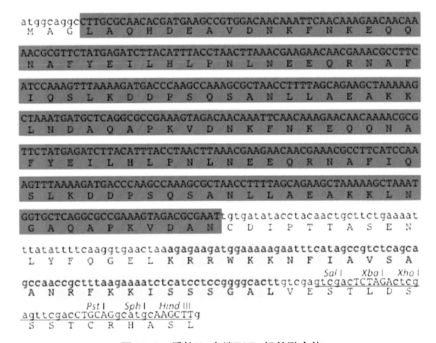

图 18-3 质粒N-末端TAP-标签融合体

如图谱所示质粒pEB587，在P_{BAD}阿拉伯糖诱导型启动子（a）和相应核苷酸和蛋白质TAP-标签序列（b）的调控下产生N-末端TAP-标签融合体。ProtA序列为深紫色，TEV蛋白酶切割位点为黄色，CBP序列为浅紫色，多克隆位点加下划线，并指示限制性酶切位点。

（宫晓炜 译）

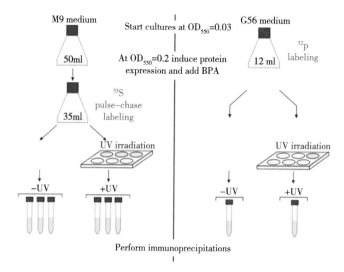

图19-1 所示实验程序的流程图

两种培养物同时进行。使用^{35}S-甲硫氨酸和^{35}S-半胱氨酸对一种培养物进行脉冲追踪标记，以标记所有新合成的蛋白质（左）。另一种培养物用^{32}P-无机磷酸盐标记，以标记细胞磷脂（右）。

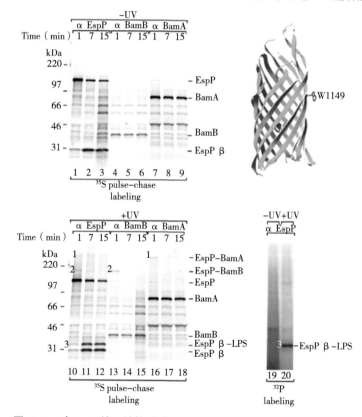

图19-2 在EspP的β结构域特定位置检测到瞬时和稳定的相互作用

RI23-1149Bpa和pDULE-pBpa转化的AD202细胞用放射性氨基酸（泳道1~18）进行脉冲追踪标记或用放射性磷酸盐标记（泳道19、20）。将每个样品分成两个等分，其中一个进行UV照射（泳道10~18和20）。使用特定的抗血清对所有样品进行免疫沉淀。图中显示了EspP（PDB：2QOM）的晶体结构，以突出W1149的位置。

（宫晓炜 译）

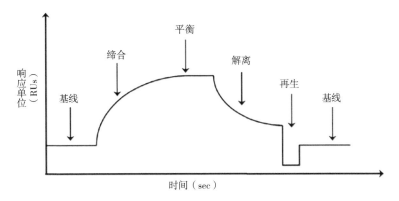

图21-1 SPR传感图的不同阶段

在配体固定后获得初始基线。在配体上注射分析物后,RUs的增加对应于缔合阶段。当缔合和解离事件的数量相等时,保持平衡水平。当停止分析物注入时,解离阶段开始:RUs的减少对应于来自配体表面的分析物的流出。再生步骤旨在完全去除分析物并达到基线水平。

$$R_{max} = \frac{MW\text{分析物}}{MW\text{配体}} \times Ri \times S$$

MW分析物:分析物的分子量。
MW配体:配体的分子量。
Ri:RUs中固定化配体的量。
S:反应的化学计量。

图21-2 理论R_{max}的确定

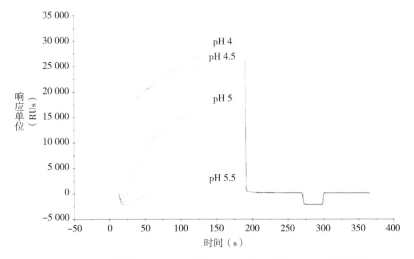

图21-3 Ⅵ型分泌系统TssE蛋白在CM5传感器表面pH值的检测

每一种测试溶液的pH值都显示在相应的声波图的顶部。

$$A + B \underset{k_{\text{off}}}{\overset{k_{\text{on}}}{\longleftrightarrow}} AB$$

$$C^A = \frac{C^A \times C^B}{C^{AB}} = \frac{k_{\text{off}}}{k_{\text{on}}}$$

C^A：分子A的浓度。
C^B：分子B的浓度。
C^{AB}：复合物AB的浓度。
K_{on}：复合物AB的关联速率常数。
K_{off}：复合物AB的解离速率常数。

图21-4 表示两个分子A和B之间相互作用的不同结合参数

$$Req = \frac{C^A \times R_{\max}}{C^A + K_D}$$

Req：平衡响应。
$R_{\max}$：分析物的最大结合能力

图21-5 在平衡和分析物浓度下使用响应单位确定K_D

$$R = \left(\frac{C^A \times R_{\max}}{C^A + K_D}\right)\left(1 - \frac{1}{e^{((k_{\text{on}} \times C^A) + k_{\text{off}}) \times t}}\right)$$

图21-6 复合物AB的结合速率k_{on}的测定

$$R = R_0 \times e^{(-k_{\text{off}} \times dt)}$$

R_0：关联结束时的响应单位

图21-7 复合物AB的解离速率k_{off}的测定

（郑福英 译）

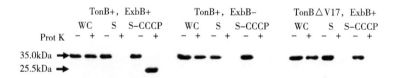

图22-1 通过差异蛋白酶K敏感性鉴定离子电化学响应性TonB构象

显示了用（+）或不用（-）蛋白酶K（"Prot K"）处理的全细胞（WC），原生质球（S）和CCCP处理的原生质球（S-CCCP）样品的蛋白质印迹。左边的样本集是由野生型菌株W3110产生的，是TonB、ExbB和ExbD的野生型。中间的样本集由W3110衍生物生成，为TonB的野生型，不表达ExbB和ExbD。右边的样品组由带有*TonB*等位基因的W3110衍生物产生，该等位基因编码无活性的TonB，缺失来自17位的缬氨酸残基，并且是ExbB和ExbD的野生型。样品在11% SDS-聚丙烯酰胺凝胶上显影，转移到聚乙烯二氟化物膜上，并用单克隆抗体4H4（针对TonB残基79～84）检测。图左侧为完整的TonB和耐蛋白酶k片段的位置和表观分子质量。该图中的一部分从Larsen等[5]转载，得到John Wiley和Sons的许可。

（郑福英 译）

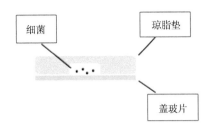

图23-1 琼脂糖垫的制备方法

（a）将约50μL琼脂糖溶液（米色）转移到显微镜载玻片上并用盖玻片（灰色）快速覆盖，然后将其轻轻均匀地压在琼脂糖上以形成均匀分布的斑块。这种简单的方法足以快速成像，但通常会导致不均匀或倾斜的琼脂糖斑块，从而降低图像质量。（b）为了确保琼脂糖斑块的表面更加平整，覆盖盖玻片可以由两个侧面盖玻片支撑。（c）双面胶带或市售粘合剂（如Gene Frame、Thermo Fisher）（波纹图案）可用于将盖玻片永久地黏附到样品上。这可以防止成像过程中的蒸发，这对于花费时间较长的实验非常有用。然而，在这种情况下应该降低样品的氧化作用。

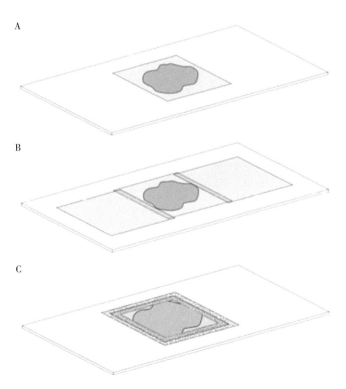

图23-2 装载到盖玻片上并用琼脂垫覆盖的细菌的示意图

（郑福英 译）

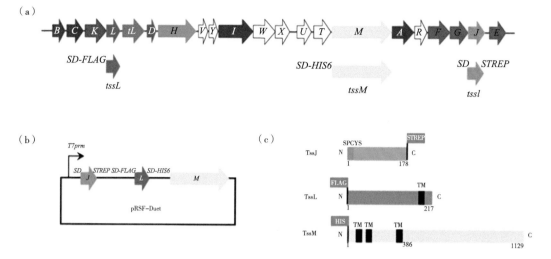

图24-1 合成的编码T6SS膜核复合物的tssJLM操纵子的克隆

（a）从聚集性sci-1 T6SS操纵子中PCR扩增tssL、tssM和tssJ基因。在每个基因的5'端加入编码优化RBS的DNA序列，为TssJ编码C端Strep-Ⅱ标签（STREP），为TssM编码N端6×His标签，以及为TssL编码Flag标签的序列也要加入。（b）在pRSF-Duet载体中克隆tssJLM人工操纵子。（c）该方案代表蛋白质构建、亚结构域边界和一些指示特征（TM跨膜区段、SP信号肽和CYS酰化半胱氨酸）。

（郑福英 译）

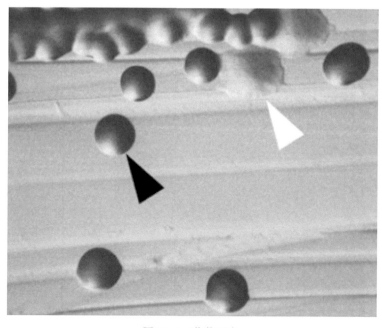

图25-1 菌落形态

将铜绿假单胞菌菌株KΔpilT涂在TSA平板上。有纤毛的菌落呈光滑的圆顶状（黑色箭头）。避免平展的菌落（白色箭头）。

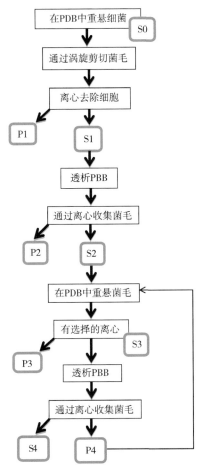

图25-2 T4aP分离的工作流程

将细菌悬浮在PDB中并通过涡旋剪切菌毛（S0）。通过离心除去细胞碎片和细菌（P1）。剩余的可溶性菌毛（S1）通过用PBB透析而聚集，并通过离心（P2）收集。出于纯化目的，将P2中的菌毛重悬于PDB（S3）中，并通过利用PBB（P4）的另一次透析进行聚集。虽然总产量有损失，也可以重复重悬和聚集的循环以改善菌毛纯度。

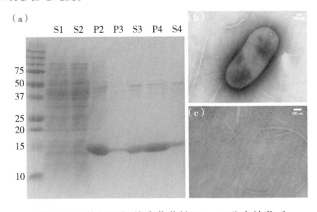

图25-3 从铜绿假单胞菌菌株K△pilT分离的菌毛

（a）在16%Tricine-SDS-PAGE凝胶中分离蛋白质（分子量标记物以kDa表示）。（b）全细胞及（c）纯化菌毛（加污染鞭毛）负染色及透射电镜分析（标尺：100nm）。

（郑福英 译）

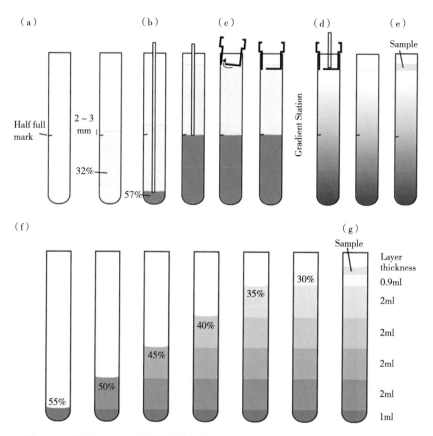

图26-1 使用Biocomp仪器的梯度管或（f~g）手工制备蔗糖梯度（a~e）

详细信息，请参阅本章3.1.4和3.1.5。

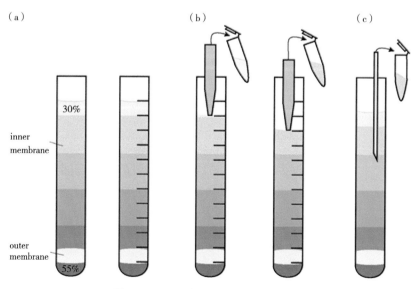

图26-2 手工收集蔗糖六步梯度馏分

（a）标记分馏的大小。（b）从顶部取分馏。（c）使用注射器取分馏，有关详细信息，请参阅3.1.5小节。

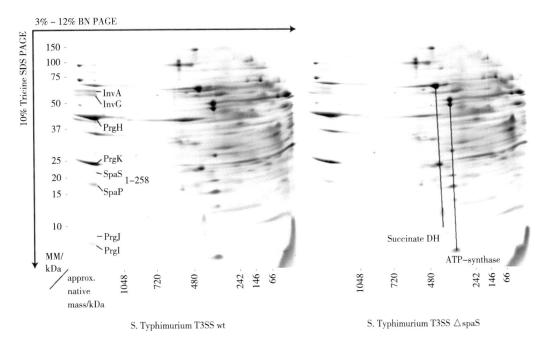

图26-3 二维BN/SDS-PAGE显示鼠伤寒沙门氏菌的考马斯亮蓝染色的内膜蛋白

左：鼠伤寒沙门氏菌野生型；指示了T3SS组件。右：鼠伤寒沙门氏菌ΔspaS突变体；显示了大量且清楚观察到的复合物。

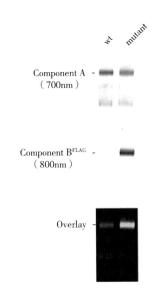

图26-4 BN PAGE分离的T3SS针状复合物的双色免疫印迹

用DDM提取野生型和突变型鼠伤寒沙门氏菌的膜的粗提物，并通过BN PAGE和免疫印迹分析。A组分可通过一抗：兔抗体和二抗DyLight 680nm（红色）山羊抗兔抗体检测。FLAG标记的B组分用一抗：抗FLAG M2小鼠单克隆抗体和二抗DyLight 800nm（绿色）山羊抗小鼠抗体检测。

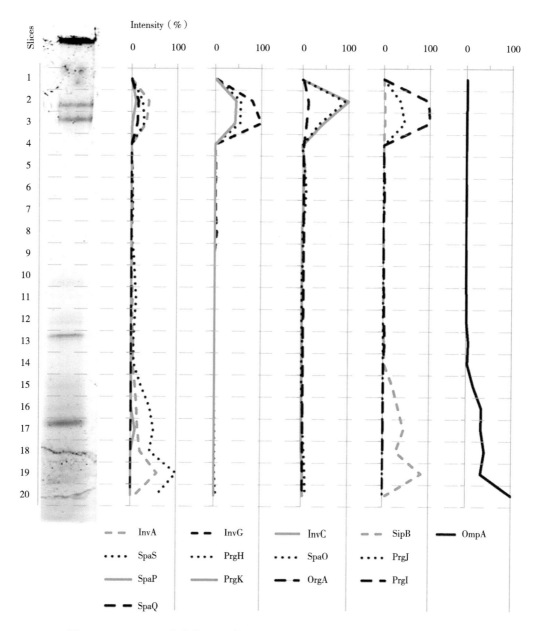

图26-5　BN PAGE分离的T3SS针状复合物的泳道分布的基于MS的分析的实例

用胶态coomassie染色的免疫沉淀的鼠伤寒沙门氏菌SPI-1 Ⅲ型分泌针状复合物的BN泳道，分成20个凝胶切片。显示了基于MS定量的蛋白质强度曲线，用于指示蛋白质。InvA、SpaS、SpaP和SpaQ是Ⅲ型分泌针状复合物输出装置的组成部分。InvG、PrgH和PrgK是针状复合物基质的组分。PrgI和PrgJ是Ⅲ型分泌系统的分泌底物，形成针状复合物的细丝结构。所有这些组分形成共免疫沉淀的稳定复合物。属于该复合物的真正组分可以从2和3条带中的相似蛋白质强度曲线推导出。显然，样品中存在大部分非针状复合物相关的SpaS和InvA（切片15～20）。InvC、SpaO和OrgA是Ⅲ型分泌系统的胞质组分，其不与针状复合物稳定结合并且在纯化期间仅共沉淀。这可以通过条带2中这些蛋白质强度的急剧峰值而不是条带3来反映。与外膜蛋白OmpA无关的蛋白可作为阴性对照。

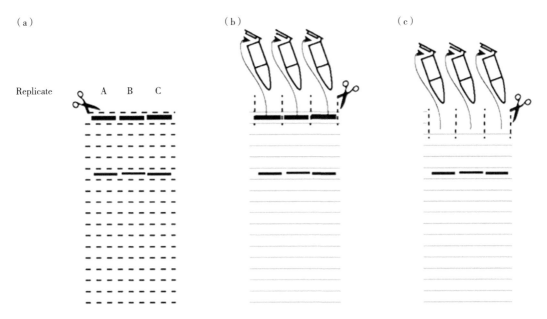

图26-6 制备用于基于MS的BN PAGE分离样品的泳道分布分析的样品

(a) 切割水平线,同时保持两侧边界完好无损。(b、c) 从上到下切割垂直线,然后在1.5mL管中收集凝胶带。有关详细信息,请参见3.3.4小节。

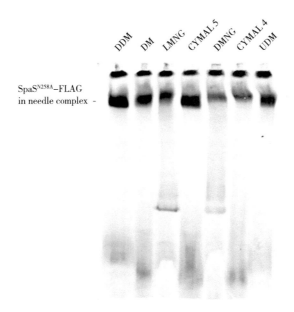

图26-7 使用一组不同的洗涤剂对T3SS针状复合物进行BN PAGE分析

用指定的洗涤剂提取鼠伤寒沙门氏菌的细胞膜,然后通过BN PAGE和免疫印迹分析。用一抗:抗FLAG M2小鼠单克隆抗体和二抗:DyLight 800nm山羊抗小鼠抗体检测FLAG标记的T3SS输出组份SpaS。检测到了Ⅲ型分泌针状复合物中的SpaS带。

(宫晓炜 译)

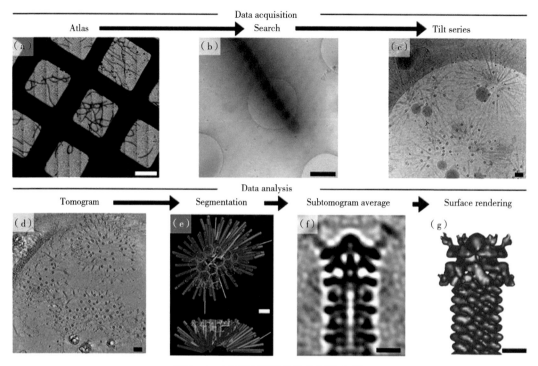

图27-1　ECT数据采集和分析的步骤

（a）覆盖几个正方形的图谱；（b）横跨碳载体的细菌细胞的搜索图像；（c）变态相关的收缩结构阵列的单独倾斜图像；（d）切片通过从c中所示的倾斜系列重建的冷冻图；（e）d中所示的断层图像分割的两个视图；（f）通过类似于d的断层图的子体积产生的平均值的切片；（g）等值面f中显示的平均值由c～e[16]的数据生成。标尺：a和b为1000nm；c～e为100nm；f和g为10nm。

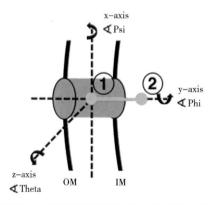

图27-2　后续X线断层照片平均的子集建模

所示为细胞包膜跨越分泌系统（OM外膜、IM内膜）的示意图。点1和2以及连接向量（绿色）表示为每个子集手动生成的模型。虚线：基于矢量方向分配给每个子体积的轴；箭头：在PEET中分配搜索角度。

（宫晓炜　译）

附　图

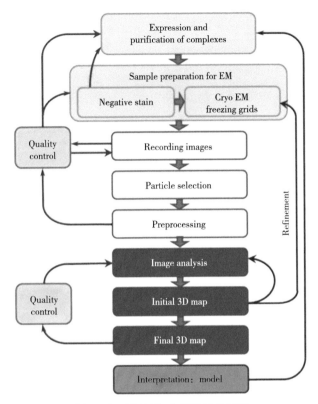

图28-1　EM结构分析的工作流程

绿色表示结构分析的实验部分。蓝色和深蓝色表示计算部分；浅蓝色表示处理的初始步骤，其中包括图像帧对齐、CTF校正、归一化和过滤。深蓝色表示后续步骤即对齐、统计分析、粒子取向的确定和初始三维重建（3D）。浅紫色表示最后一步即对所获得图像的解释。

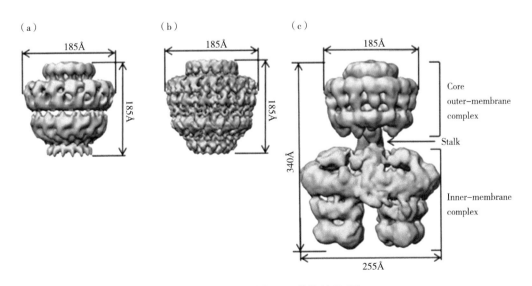

图28-2　不同T4SS结构的EM图

（a）核心外膜复合物的Cryo-EM结构，分辨率为15Å。（b）核心外膜复合物的Cryo-EM结构，分辨率为12.4Å。（c）几乎完整的T4SS复合物（负染色）的结构，总分辨率大约为20Å。

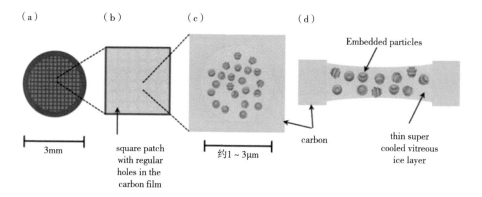

图28-3　Cryo-EM样品制备

（a）3mm铜网格覆盖有多孔碳膜。（b）方形贴片的放大图像，显示碳中的微观孔。（c）含有一层带有蛋白质分子的玻璃化冰单个孔的放大图像。（d）具有嵌入冰中的颗粒的孔的横截面。

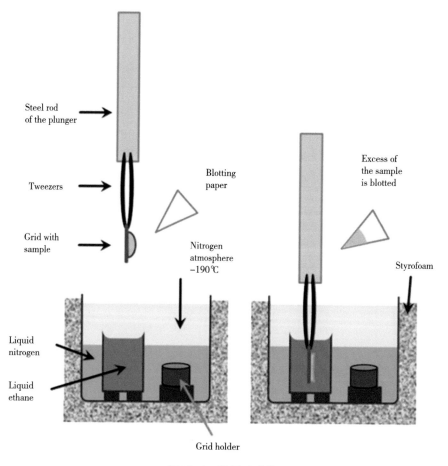

图28-4　样品玻璃化

左图：带有样品的网格用镊子固定；右图：在吸去过量样品后，将网格插入装有液体乙烷的容器中。容器的顶层应浸入温度略高于液氮温度的氮气中。然后将网格转移到网格支架中。必须在不将电网从氮气中取出的情况下进行转移。

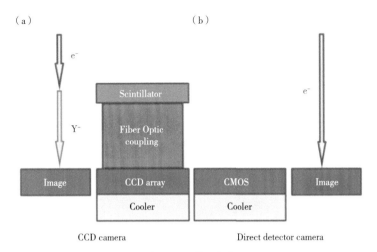

图28-5　数码相机

（a）在基于CCD的照相机中，电子撞击闪烁体，产生光，光被光纤部分捕获，并被导向冷却的CCD芯片。（b）在直接探测器（双透面板）中，主要基于CMOS技术的有源像素传感器能够捕获并直接探测入射电子。

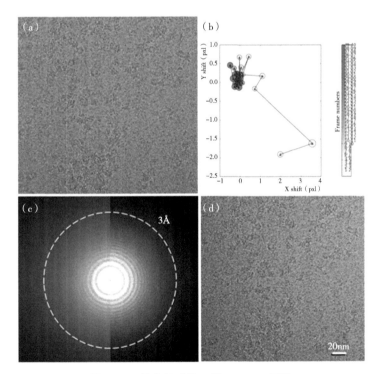

图28-6　具有运动校正的Cryo-EM图像

（a）玻璃化T4SS颗粒的代表性Cryo-EM图像。（b）在帧的X和Y方向上的运动轨迹。（c）左：来自没有运动校正的原始记录帧之和的功率谱；右：运动校正后记录帧总和的功率谱。（d）根据（b）中所示的确定的移位移动的记录帧的总和。在这些图像中黑色为蛋白质。

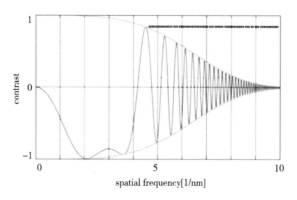

图28-7　具有包络函数的CTF

蓝色虚线：完美显微镜中所有频率的幅度；绿线：包络函数对CTF（红色）的影响导致高空间频率的抑制。

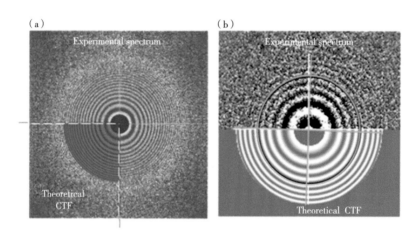

图28-8　CTF参数的评估

（a）理论计算的CTF（左下象限）与实验光谱中的CTF的比较。为了准确地确定CTF，来自两个图像部分的Thon环应该准确匹配。（b）识别散光轴，其叠加在实际观察功率谱的Thon环上并与理论光谱比较。这里显示的显微照片的光谱表明有一个小的散光，约2%，椭圆的轴稍微倾斜，以浅蓝色显示。

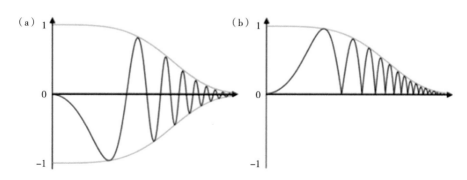

图28-9　CTF校正

（a）CTF根据频率摆动从负面到正面的对比度变化。只有当CTF越过零线时，信息才会丢失。（b）未校正CTF的负叶被翻转为正（通过相位翻转校正CTF）。可以通过以不同的散焦水平收集图像来恢复丢失的信息，所述散焦水平用信息填充这些零区域。

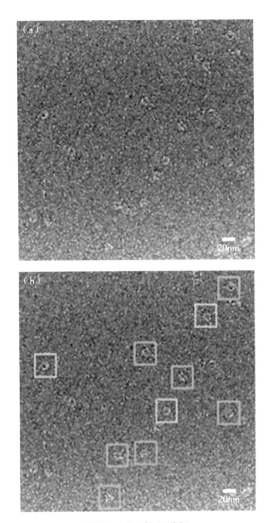

图28-10 粒子选择

（a）T4SS核心外膜复合物的Cryo-EM显微照片。（b）黄色方块表示的粒子的最终视角图；蓝色方块为粒子的侧视图。

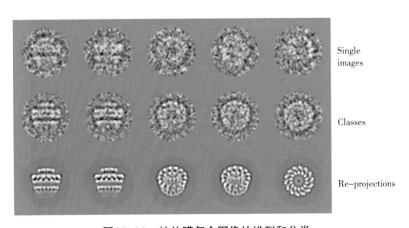

图28-11 核外膜复合图像的排列和分类

上面一排：核心外膜复合体的代表性图像；中间一排：几乎相同的图像类别平均数；底部一排：对应于最终三维模型的图像。

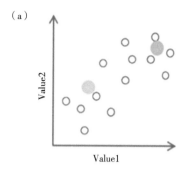

Step 1. K=2. Two seeds are selected randomly

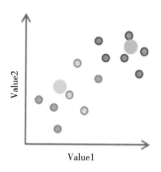

Step 2. Elements are assigned to the classes of the nearest seed

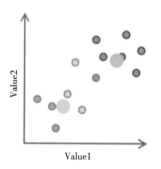

Step 3. The averages are recalculated based on the assignments in step 2. Steps 2 and 3 are reiterated

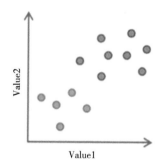

Step 4. Final classes

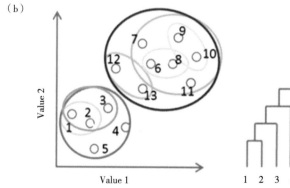
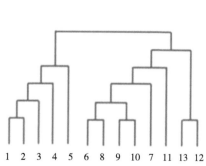

图28-12　图像分类原理

（a）K-means分类：实验数据由空心圆表示，其特征参数为v1和v2；初始随机选择的种子以浅蓝色和粉红色显示；分类步骤的结果以相应的圆圈显示。（b）分级分类：左侧面板显示了平面内点的位置及其连续形成的类；右侧面板显示分类树。

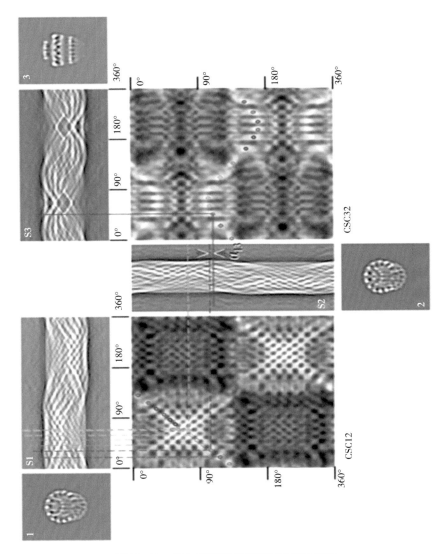

图28-13 三个投影的正弦图和正相关函数

样本对象是T4SS核心外膜复合体的模型，具有14倍对称性。对应于投影的图像编号为1~3。S1、S2和S3（正弦图）是相应投影1~3的一维投影的集合。CSC12和CSC32分别对应于投影1和2以及投影3和2之间的交叉正弦图相关（CSC）函数。CSC函数的每个点表示来自两个正弦图的一对线的相关系数。每对图像之间有14条共同线，因为对象具有14倍对称性。最高的相关性（彩虹圆圈）指向公共线的位置。它们中的一些由突起2和1之间的虚线示出，而实线红线示出了突起2和3之间的一条公共线（另一条未示出）。每个CSC功能都将所有峰值加倍，因为从180°到360°的投影反映了从0°到180°的投影。公共线（红色实线和虚线）之间的角距离表示突起1和3之间的角度α_{13}。

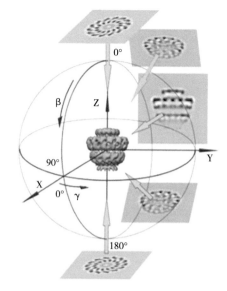

图28-14　现实空间中的3D重建

显示了核心外膜复合物的不同等级平均值并且位于欧拉球范围。每个类平均值在实际空间中沿其指定的欧拉角反投影。每个2D类平均值的密度通过3D空间被拉伸为光线，并且交叉光线的交叉点，将总和定义整个对象的3D电子密度。

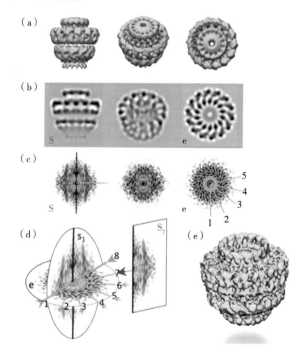

图28-15　使用类别（图像）的傅里叶变换（FT）进行重建

（a）在支撑膜上的不同孔中观察到T4SS核心外膜复合物（14倍对称）。（b）这些粒子的投影等同于电子束方向上的EM图像。（c）来自b中所示预测的FT。s：侧视图的FT；e：最终视图的FT。相应的图像显示在b中。（d）在倒易空间中共享至少一条公共线的一对2D变换。侧视图投影和端视图（e）投影之间的公共线用绿线表示；紫色线：侧视图之间的公共线（如s1和s7所示）。公共线对之间的角度确定相对欧拉角方向。（e）2D组合变换的逆傅里叶变换形成的，产生改进的实空间结构，称为电子密度。

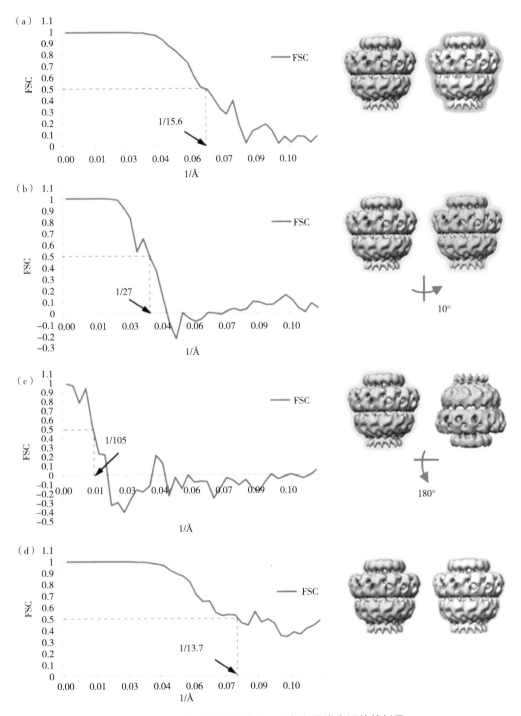

图28-16 傅里叶壳关联（FSC）和分辨率评估的例子

（a）两个独立的3D结构的FSC曲率已经很好地对齐，并且用松散的软掩模掩盖，该掩模在粒子图像周围显示为宽晕。评估为0.5级的分辨率为15.6Å。（b）这里，第二结构围绕旋转轴旋转10°。虽然两个结构之间的整体形状仍然很好，但小的细节不在寄存器中。它反映在FSC曲线中，其下降得更快，0.5阈值处的分辨率仅表示27Å，对应于结构中主要区域的大小。（c）结构不一致。FSC在较低频率下甚至下降更早，表明整体尺寸只有一致性。（d）当应用相同的紧密掩模时，两个3D结构之间的FSC。高频率的增加表明施加在结构上的掩模之间的相关性，此处的分辨率被高估。

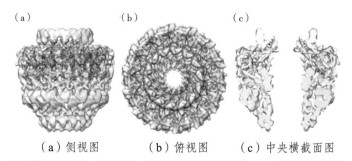

（a）侧视图　　　（b）俯视图　　　（c）中央横截面图

图28-17　原子模型（PDB：3JQO）与核心外膜复合物的cryo-EM图（EMD-2232）的拟合

（陈启伟 译）

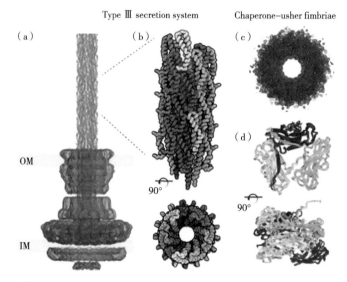

图29-1　基于SSNMR的方法对Ⅲ型分泌系统的针和1型菌毛的结构研究进行概述

（a）基于福氏志贺菌针（登录号EMD-5352[2]）和鼠伤寒沙门氏菌针复合物（登录号EMD-1875[70]）的cryo-EM图谱的Ⅲ型分泌系统注射体的示意图。（b）基于SSNMR的鼠伤寒沙门氏菌菌针丝的原子模型[1]，侧视图和俯视图，登录号PDB-2LPZ。（c）通过cryo-EM获得的chaperone–usher菌毛结构的顶视图（登录号EMD-3222[6]）。（d）基于SSNMR的原子分辨率模型（登录号PDB-2N7H[5]）。

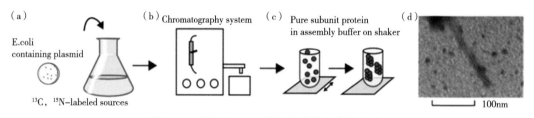

图29-2　用于SSNMR分析的体外细菌丝重建

（a）标记^{13}C、^{15}N的蛋白质亚基在基本培养基中的表达。必须从琼脂平板中选择一个转化的大肠杆菌菌落以接种预培养物，其进一步用于接种主培养物。（b）破坏细胞并收获蛋白质亚基并在适当的纯化柱系统上纯化。（c）将纯蛋白质亚基在装配缓冲液中浓缩（或稀释）至合理的装配浓度，应在最佳pH值和盐浓度下仔细测试和优化。装配条件通常在缓慢摇动下是最佳的。然后子单元通过组装（在此表示）采用其原始构象。（d）细丝形成应通过透射电子显微镜观察，如图所示，蛋白质细丝（亚基大小为14kDa）。

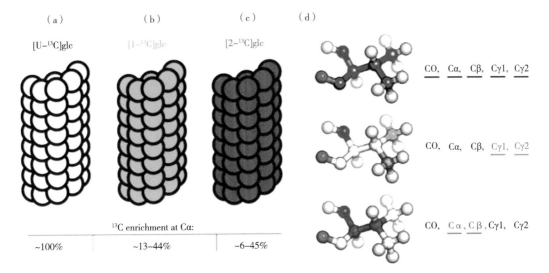

图29-3 蛋白质亚基同位素标记的策略

均匀相标记方案，基于（a）均匀^{13}C标记的亚基（白色）、（b）选择性[1-^{13}C]-葡萄糖标记的亚基（绿色）和（C）选择性[2-^{13}C]-葡萄糖标记亚单位（粉红色）。[1-^{13}C]-和[2-^{13}C]-葡萄糖标记方案是均一的，即所有蛋白质亚基都用相同的方案标记，尽管同位素标记的碳位置是具有选择性的。（d）每种标记方案的氨基酸缬氨酸的标记模式。根据情况a～c，同位素标记的原子位置加下划线并着色。

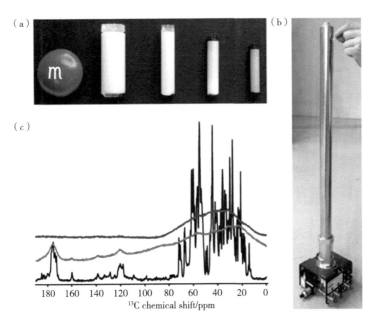

图29-4 固态实验装置

（a）SSNMR MAS转子，直径为7、4、3.2、2.5mm。（b）布鲁克三重共振^{1}H/^{13}C/^{15}N MAS 3.2mm探头。（c）在MAS条件下记录的蛋白质丝的碳检测的^{1}H-^{13}C CP光谱，在采集期间具有^{1}H去耦（SPINAL64）[54]（黑色），在MAS条件下没有^{1}H去耦（红色），没有MAS（蓝色）。

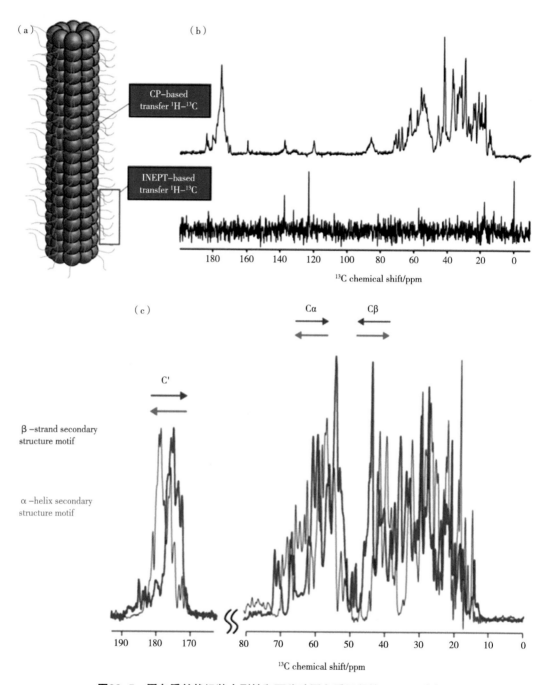

图29-5 蛋白质丝状组装中刚性和可移动蛋白质区段的SSNMR表征

（a）细菌丝的示意图，包括刚性和可移动的蛋白片段。（b）基于交叉极化（CP）的实验揭示，有助于组件刚性核心的残留物。基于INEPT的实验探测了移动片段的存在，在亚基大小为14kDa的蛋白质丝上进行了说明。（c）C_α、C_β和羧基C'原子的化学位移值对二级结构敏感，化学位移分布表示二级结构组成；在含有β-链（蓝色）和α-螺旋（红色）构象的两种蛋白质细丝（两种亚基大小均为9kDa）的1D CP光谱上进行说明。

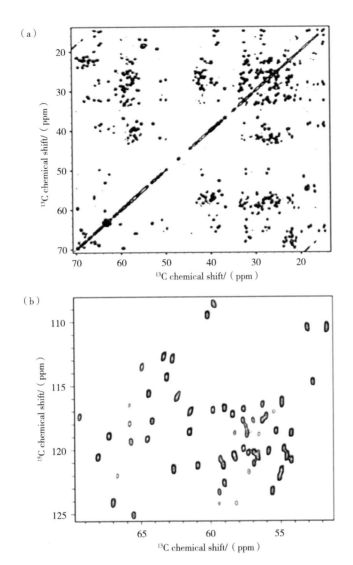

图29-6 指纹SSNMR实验用于细菌丝状附属物的结构研究

（a）以短混合时间记录的2D ^{13}C-^{13}C PDSD光谱，以观察残基间相关性。（b）^{15}N-^{13}Cα二维谱以探测残基骨架相关性。这些实验在鼠伤寒沙门氏菌Ⅲ型分泌系统菌针上进行了说明[1]。

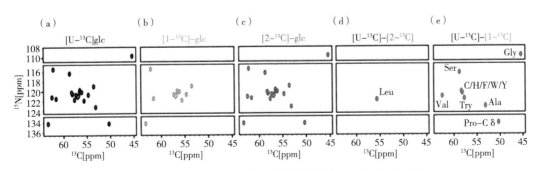

图29-7 每种氨基酸三种不同标记方案的2D NCA光谱比较

[U-^{13}C]-glc（a），1-^{13}C]-glc（b）和[2-^{13}C]-glc（C）（d）[U-^{13}C]和[2-^{13}C]-glc的光谱比较导致亮氨酸相关性的鉴定。（e）[U-^{13}C]和[1-^{13}C]-glc的光谱比较可以确定几种氨基酸的相关性。

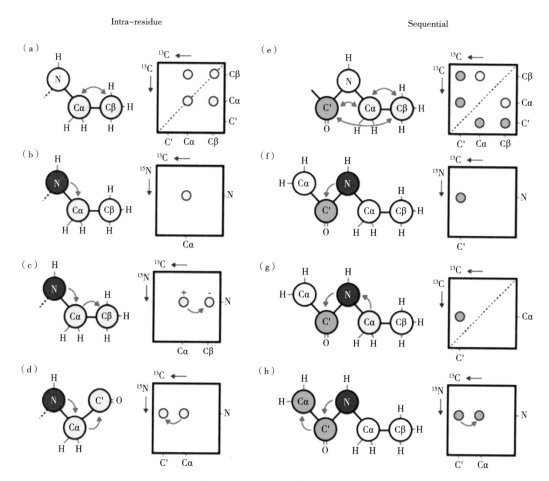

图29-8 SSNMR共振分配策略

通过碳-硼和氮-硼原子之间的连接建立残留（a~d）和残基（e~h）的相关性。（a）所有残留 $^{13}C-^{13}C$ 相关性（短混合时间）的C-C PDSD实验。（b）用于残留 $^{15}N-^{13}C\alpha$ 相关性的N-C特异性CP。（c）N-（Cα）-Cβ（ $^{15}N-^{13}C\alpha$ 特异性-CP，DREAM）用于残留 $^{15}N-^{13}C\alpha-^{13}C\beta$ 的相关性。（d）N-（Cα）-C'($^{15}N-^{13}C\alpha$ 特异性-CP，MIRROR）用于抑制 $^{15}N-^{13}C\alpha-13C$ 的相关性。（e）C-C PDSD实验主要用于抑制和连接 $^{13}C-^{13}C$ 的相关性（中间混合时间）。（f）N-C'($^{15}N-^{13}C$ 特异性-CP）对于连续的 $^{15}N-^{13}C$ 的相关性。（g）顺序 $^{13}C\alpha-^{13}C$ 相关的Cα-（N）-C（ $^{13}C\alpha-^{15}N$ 特异性CP， $^{15}N-^{13}C$ 特异性CP）。（h）连续 $^{15}N-^{13}C-^{13}C\alpha$ 相关的N-（C'）-Cα（ $^{15}N-^{13}C$ 特异性CP，MIRROR）。

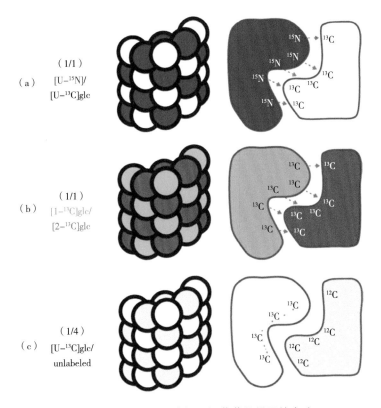

图29-9 异质同位素标记细菌菌丝样品的方法

（a）均匀^{13}C（蓝色）和均匀^{15}N（白色）标记亚基的等摩尔混合物；（b）[1-^{13}C]-葡萄糖（绿色）和[2-^{13}C]-葡萄糖（红色）标记亚基的等摩尔混合物；（c）未标记亚基（白色和黄色）中^{13}C标记亚基的稀释混合物（1/4）。

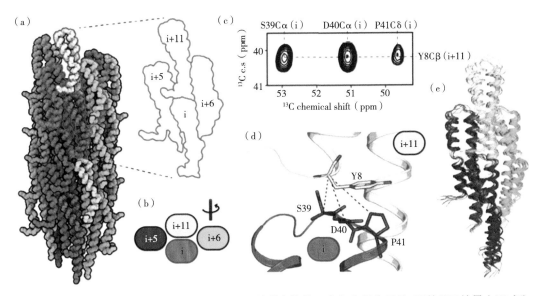

图29-10 （a、b）T3SS分泌系统针状长丝及其基本构件，由包含所有亚基-亚基界面的最小不对称单元组成。（c）分子间SSNMR约束的检测；（d）亚基-亚基接口的距离限制；（e）T3SS分泌系统构件的SSNMR原子结构

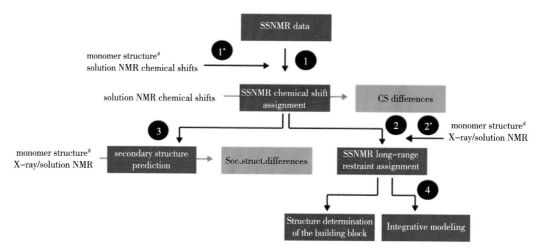

图29-11　基于SSNMR的综合结构确定蛋白结构过程

对不同的任务进行编号，表示可以集成的数据。如果有必要的数据，绿色箭头和下划线突出显示中间结果。第1步：分配过程；第2步：SSNMR数据中远程接触的分配过程；第3步：二级结构确定；第4步：结构建模与从生物物理技术获得的不同结构数据的整合。#从X射线晶体学或溶液NMR获得的单体结构。

（陈启伟　译）

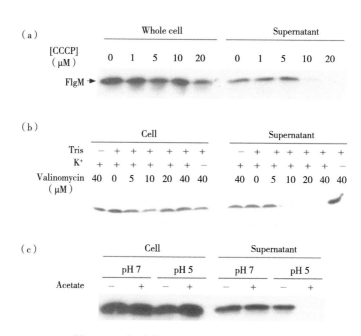

图30-1　（a）抑制pmf对FlgM输出的影响

通过添加10μM解偶联剂羰基氰化物间氯苯腙（CCCP）抑制FlgM分泌，并通过在菌株TH10874（阿拉伯糖诱导型$flgM$）中用20μM CCCP处理完全消除。细胞质FlgM水平保持不变。（b）抑制pmf的ΔΨ组分对FlgM输出的影响。在K^+存在下加入缬氨霉素来抑制FlgM分泌。用120mM Tris-HCl预处理细胞，使外膜透化缬氨霉素，如图所示。（c）抑制ΔpH对FlgM输出的影响。通过加入34mM乙酸钾抑制在pH值为5下生长培养物的FlgM分泌。改编自Macmillan Publishers Ltd许可：Nature[6]，2008。

（陈启伟　译）

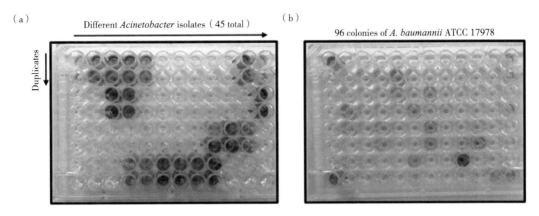

图32-1 （a）利用Hcp ELISA同时筛选某一物种的多个分离株

本文对45株不同不动杆菌的Hcp分泌株进行了重复筛选。结果表明，在一个特定的属内，T6SS活性在不同物种间存在差异。（b）筛选单一菌株，鲍曼尼菌株ATCC 17978，用于Hcp分泌。采用单菌落接种法进行Hcp酶联免疫吸附试验。对于该菌株，质粒编码T6SS的抑制因子，但在没有选择[12]的情况下可在培养过程中丢失。

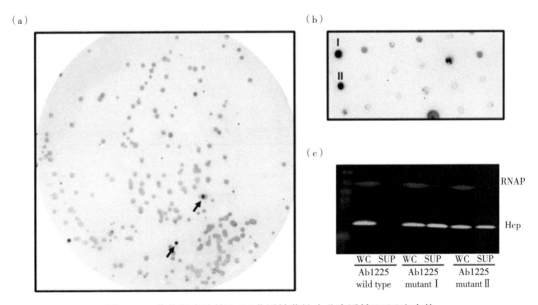

图32-2 菌落印迹法从T6SS非活性菌株中分离活性T6SS突变体

（a）用于筛选鲍曼氏A. baumannii菌株1225转座子突变体的Hcp菌落印迹实验，该菌株具有失活的T6SS，在正常实验室条件下不分泌Hcp。箭头表示用抗hcp抗体检测的单个菌落的强信号。（b）鲍曼氏转座子突变体表面Hcp菌落印迹。菌落Ⅰ和Ⅱ表现出强烈的抗hcp信号。（c）Western blot检测各组Ⅰ、Ⅱ菌落全细胞（WC）Hcp表达及上清液（SUP）分泌情况（见图b）。野生型鲍曼氏菌1225表达但不分泌Hcp，作为阴性对照。

（宫晓炜 译）

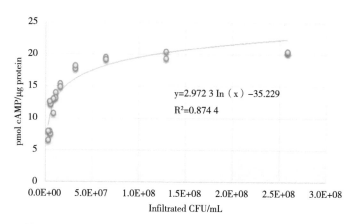

图33-1 基于cya报告子的T3SS易位分析：cAMP积累与接种水平的非线性关系

悬浮转化了携带avrpto1 Cya表达质粒pCPP5312的Pto菌DC3000，并渗透通过两片烟叶中，使浓度达到$2.6 \times 10^8 \sim 2.7 \times 10^6$cfu/mL。接种后6h，每片烟叶收获直径为1cm的叶盘，测定可溶性pmol cAMP/μg蛋白。用88个处理过的叶片圆盘计算平均可溶性蛋白浓度pmol cAMP/μg蛋白。

（宫晓炜 译）

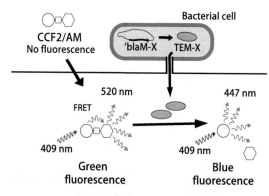

图34-1 评价真核细胞中效应蛋白易位的TEM-1报告系统示意图

当被动进入真核细胞时，非荧光酯化CCF2/AM（或CCF4/AM）底物被细胞酯酶迅速转化为带电荧光CCF2。409nm处香豆素部分（圆）的激发导致荧光能量转移（FRET）到荧光素部分（Hexagon），后者发出绿色荧光。在520nm处发出E信号。将融合到TEM-1的效应物注射到承载CCF2的细胞中，可诱导CCF2β-内酰胺环（正方形）的催化裂解，从而破坏FRET。这使得荧光从绿色到蓝色的发射很容易被检测到和测量到。

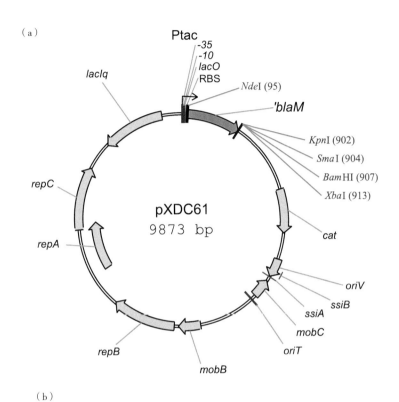

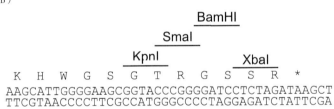

图34-2 表达β-内酰胺酶TEM效应物融合蛋白的质粒（pxDC61）图谱

（a）pxDC61质粒系衍生自广泛宿主范围质粒RSF1010之可移动（oriT）质粒。质粒具有氯霉素抗性，'blaM基因编码的是TEM-1 β-内酰胺酶的成熟形式，该酶失去了其N端分泌信号，由IPTG诱导启动子控制。一个多克隆位点（KpnI、SmaI、BamHI、XbaI）被放置在'blaM基因的3'端。（b）用于效应基因的多克隆位点示意图。

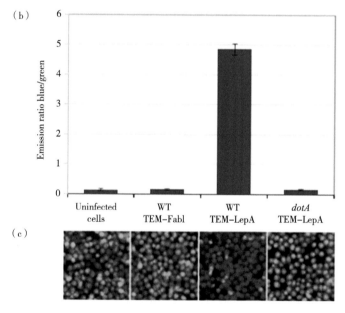

图34-3 易位试验的经典结果

（a）原始数据，通过嗜肺菌株野生型（WT）或ΔDotA缺失株中已知Dot/Icm LepA效应物和非分泌胞质蛋白FabI获取减去空白的荧光信号（相对荧光单位，RFU）和荧光比值（发射460nm/530nm）。每个菌株的分泌物测定三份。用配备单色仪和顶部荧光读取仪Tecan M200进行荧光测量。测量程序包括激发波长为405nm、发射波长为460nm和530nm的荧光读数，增益设置为135。（b）图形表示LepA和FabI的平均分泌比和相应的标准偏差。依据两种荧光测量增益的方法，这个比值可以发生显著变化。通过设置未感染细胞460/530的比例为1，将不同增益或不同平板获得的数据进行均一化。（c）典型易位试验的荧光图像。在405nm激发下，将460nm和530nm下捕获的图像进行合并。

（宫晓炜 译）

附　图

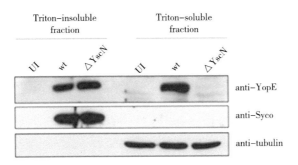

图35-1　小肠结肠炎耶尔森菌感染RAW 264.7巨噬细胞样细胞时的效应蛋白易位

RAW 264.7细胞感染野生型（WT）或T3SS缺陷型（YscN突变体）大肠杆菌菌株。如2.1和3.1小节所述，制备了未感染细胞（UI）和WT或YscN突变菌（ΔYscN）感染的细胞的Triton可溶性部分和非溶部分。按照2.2、2.3和3.3小节所述，通过SDS-PAGE和免疫印迹分析样品。Yope是一种大肠杆菌效应蛋白；Syco是一种大肠杆菌T3S伴侣蛋白[15]，作为对照，可验证Triton可溶性部分是否存在显著的细菌交叉污染（并作为Triton不溶部分加样时的对照）；Tubulin是一种宿主细胞蛋白，用作Triton可溶性部分加样时的对照。

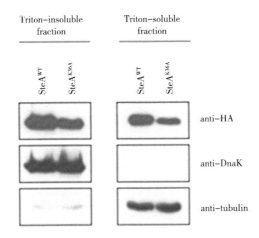

图35-2　HELA细胞感染肠沙门氏菌（沙门氏菌）引起的效应蛋白易位

HELA细胞被鼠伤寒链球菌脂肪突变体感染，该突变体携带编码C末端2×HA标签标记的野生型脂肪（脂肪-2HA）或36位赖氨酸残基替代丙氨酸（脂肪36a-2HA）的突变脂肪。按2.1和3.2小节（见注释29）制备了两种菌株感染细胞的Triton可溶和不溶性部分。按照2.2、2.3和3.3小节所述，采用SDS-PAGE和免疫印迹法对样品进行分析。SteA是一种沙门氏菌效应蛋白[10, 14]，DnaK是一种细菌分子伴侣，作为对照，可验证Triton可溶性部分是否存在显著的细菌交叉污染（并作为Triton不溶部分加样时的对照）；Tubulin是一种宿主细胞蛋白，用作Triton可溶性部分加样时的对照。注意检测Triton不溶性部分中微管蛋白的残留水平。

（宫晓炜　译）

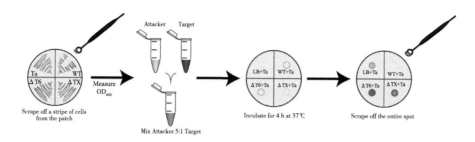

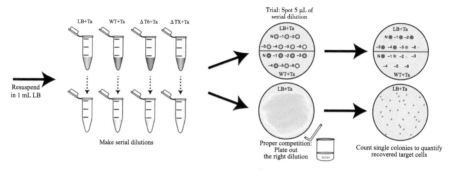

图36-1 抗菌活性测定示意图

在规定的培养期，在规定的温度下，刮除共培养点（LB+靶标，绿色；野生型+靶标，蓝色；ΔT6SS+靶标，红色；Δ毒素+靶标，灰色），重新悬浮细胞并制备系列稀释液。在初步试验中，这些纯（n）的10^{-6}的稀释液被发现在一个琼脂板上，该板上补充了对目标生长有选择性的抗生素。孵育后，用试验平板的目标回收率的估算来确定稀释度，这将在实际实验中为每个平板提供几十个单菌落。在实际实验中，将适当稀释的共培养液用玻璃涂布器铺在选择性平板上，并在过夜培养后计数菌落；重复实验提供了完全定量的数据。有关详细信息，请参阅文本。

	R1	R2	R3	R4	Mean	SEM
LB	6.0E+08				6.0E+08	
WT	1.1E+04				1.1E+04	
ΔT6SS	4.5E+08				4.5E+08	

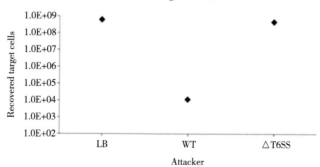

图36-2 抗菌活性测定数据的图形表示

上面：表中显示了经典的测定（正文中的示例）产生的每个共培养菌的菌落形成单位数。R1表示第一次复制，因此平均值对应于一个重复，没有标准偏差（SEM）；但是，在正规的实验中，至少应进行四次重复，并呈现平均值±SEM。下面：使用Microsoft Excel生成带有表中提供的数据图形。Y轴显示复苏的目标菌数，以每个共培养斑点的菌落形成单位数表示，X轴显示攻击菌株：对照（LB）、野生型菌株（WT）和T6SS突变体菌株（ΔT6SS）。

（宫晓炜 译）